Epistemology, Science and Cognition

Hyderabad Studies in Philosophy

1. Philosophy of Wittgenstein: Indian Responses; by R.C. Pradhan (ISBN 978-81-86921-15-9)
2. The Philosophy of Language in Classical Indian Tradition; by K.S. Prasad (ISBN 978-81-86921-20-3)
3. Indian Philosophy of Religion; by A. Ramamurty (ISBN 978-81-86921-22-7)
4. Language and Mind, Vol. 1: A Western Perspective; edited by R.C. Pradhan (ISBN 978-81-86921-34-0)
5. Language and Mind, Vol. 2: The Classical Indian Perspective; edited by K.S. Prasad (ISBN 978-81-86921-45-6)
6. Epistemology, Science and Cognition; edited by Prajit K. Basu and S.G. Kulkarni (ISBN 978-81-86921-54-8)
7. Communicate Self and Other Essays: Philosophy of Ramachandra Gandhi; edited by R. Raghuramaraju (forthcoming)
8. Language: An Instrument of Knowing; edited by Ananda V. Wazalwar (forthcoming)
9. Mind and Meaning; edited by Prajit K. Basu (forthcoming)

Hyderabad Studies in Philosophy No. 6

Epistemology, Science and Cognition

Editors

Prajit K. Basu
S.G. Kulkarni

Decent Books
New Delhi

Cataloging in Publication Data — DK
[Courtsey: D.K. Agencies (P) Ltd. <docinfo@dkagencies.com>]

Epistemology, science and cognition / editors, Prajit Basu, S.G. Kulkarni.
p. cm. – (Hyderabad studies in philosophy ; no. 6)
Papers presented at two national seminars on Language science and cognition and Epistemology and cognition held at Hyderabad.
Includes bibliographical references.
Includes index.
ISBN 13: 9788186921548 ISBN 10: 8186921540

1. Knowledge, Theory of – Congresses. 2. Knowledge, Theory of (Hinduism) – Congresses. 3. Cognitive science – Philosophy – Congresses. 4. Cognition – Philosophy – Congresses. I. Basu, Prajit K. II. Kulkarni, S.G. III. Series: Hyderabad studies in philosophy ; no. 6.

DDC 121 22

ISBN 13: 978-81-86921-54-8 ISBN 10: 81-86921-54-0
First published in India, 2011

Published and printed by:
Decent Books
D-36/A-1, Mohan Garden
Najafgarh Road, New Delhi - 110 059
Ph. (011) 2535 3435

Distributed by:
D.K. Printworld (P) Ltd.
Regd. office : *"Srikunj,"* F-52, Bali Nagar
Near: Ramesh Nagar Metro Station
New Delhi - 110 015
Phones : (011) 2545 3975, 2546 6019; *Fax* : (011) 2546 5926
E-mail : indology@dkprintworld.com
Web : www.dkprintworld.com

Acknowledgement

THE present volume entitled *Epistemology, Science and Cognition* is the sixth under the series **Hyderabad Studies in Philosophy**. We are grateful to the University Grants Commission, India, for providing the necessary grant for the publication of this volume under the Special Assistance Programme (DSA-SAP). We are also thankful to the authors for sending us the revised versions of their papers and for their consent to include their papers in this volume.

We take this opportunity to place on record our gratitude to our Vice-Chancellor and the Chairman of the DSA-SAP Advisory Committee, Professor Seyed E. Hasnain for providing all necessary help in undertaking this publication. We are indebted to Professor Amitabha Dasgupta, Coordinator of the DSA-SAP, for his keen interest and assistance in the publication of this volume and for extending all help in organizing the seminar. We are also extremely grateful to Professor R.C. Pradhan, Professor A. Raghuramaraju, Dr K.S. Prasad, Dr C.B. Varma, A.V. Wazalwar, and A. Joshi for their help and encouragement. The university administration and especially the department office have provided the much needed support. We also deeply appreciate the students of the Department of Philosophy who have always demanded a vigorous academic environment in the department.

We also thank Susheel K. Mittal and Rajendra Agarwal of Decent Books, New Delhi, for undertaking this publication, and also for being very patient.

University of Hyderabad

Prajit K. Basu
S.G. Kulkarni

Contents

PART II
Cognition and Science

PART III
Cognition and Mind

Introduction

The present volume titled, *Epistemology, Science and Cognition*, is an attempt to bring out some of the important issues that arise at the interface of epistemology, philosophical studies of science, and the present research in the area of cognitive science. There are fifteen papers in this edited volume. These papers were presented in two National Seminars titled "Language, Science and Cognition" and "Epistemology and Cognition," organized by the Department of Philosophy, University of Hyderabad, under the UGC-DSA-SAP (2004-2009). There were twenty-two papers presented in these seminars. Fifteen papers have been included in this volume since either the other papers were somewhat less topical given our interest or the papers failed to reach us in time. Each paper was presented and thoroughly discussed after the presentation during the seminar and the authors subsequently revised the papers based on the discussion.

One of the consequences of the post-positivist philosophy of science is the emergence of the cognitivist approach in philosophy of science. The philosophers of science who have adopted this approach trace some central concepts of the cognitivist framework to the ideas of T.S. Kuhn and even P.K. Feyerabend. Some questions in connection with this new development are: (a) Does the cognitivist approach help us to transcend or re-configure the normativist-naturalist dichotomy?; (b) How far can the cognitivist approach develop an analytical schema to specify the relative importance of cognitive factors in providing an epistemological account of both the statics and the dynamics of science?; (c) What is the significance and what are the limits of modelling as the basis of scientific reasoning?; and (d) What light does the new approach shed on the rules of scientific and day-to-day reasoning?

The realization of the simplistic character of the positivist construal of the vocabulary of science led to a detailed analysis of the linguistic apparatus of science, especially in the light of seminal developments in Philosophy of Language and Theoretical Linguistics. The linguistic turn in the post positivist philosophy of science resulted in radical implications for our notion of meaning, reference and truth in the context of science as a cognitive enterprise. The incommensurability thesis and responses to it are of immense significance in this connection, especially because, among many other things, inter-theoretic translation has been shown to be problematic.

The recognition, following A. Goldman's seminal work, that psychology and cognitive science are essential to the enterprise of epistemology and B.K. Matilal's insight as to how certain strands of Indian philosophy may be profitably explored to sharpen issues raised by Goldman's suggestion open up interesting questions in epistemology. The attempt is to displace the primary locus of evaluation, in epistemology, from claims of knowledge to processes of belief formation in individual knowers. The epistemic evaluation is thus based on psychology of claim makers. Psychological processes are proper objects of epistemic evaluation and the question that needs to be asked is as to how these processes are to be evaluated. Given the physical limitations that a cognitive agent has in terms of not being logically omniscient, and the finite number of beliefs possessed by the agent, the time constraint within which beliefs must be processed to achieve her goal, etc., the reliability of the methods employed by the cognitive architecture of the agent assumes importance. And the notion of reasoning and rationality become tied to the central question of (and a plausible answer to) epistemology.

In the light of the above-mentioned issues the papers in this volume are organized and need to be read and perused.

In Part I, there are eight papers addressing different aspects related to epistemology and the role of an account of cognition in clarifying some of the important questions in epistemology and thereby suggesting novel solutions to those questions. These questions relate to frontal challenges made by skeptics concerning the possibility of knowledge, or the plausibility of the principle of knowledge being closed under logical implication adduced by the skeptics to argue for their claims. The attempted responses from the Western and Indian traditions are taken up by two authors in this part. An account of epistemology brings in invariably the question of rationality. There is a serious and novel attempt to view this question through the lens of sociality and historicality. There are three papers which consider the cognitivist paradigm of epistemology. The paper "Six Responses to Skepticism" by Sandhya Basu summarizes the problem posed by the skeptic, interpreted through the understanding of knowledge as indefeasibly justified true belief, as to whether knowledge at all is possible. The paper keeps its focus on the problem of knowledge about the external world. The author contends that the tall claim of philosophical skepticism that it cannot be confuted by experience is a vacuous one and its contention that it can fly high since it cannot be grounded by any refutation is untenable. She also mentions the futility of easy solutions to the problem of skepticism such as those suggested by A.J. Ayer (who admits the serious limitations of his argument) and G. Ryle. She then takes up G.E. Moore's challenge to skepticism and finds Wittgenstein's response to it to be wanting. After looking at the anti-skeptical responses by R.M. Chisholm and P.F. Strawson, she sympathetically considers Davidson's position and highlights its significance for her position that we have as much right to choose an anti-skeptical way of life as the skeptical one. Thus, the skeptic may not have been defeated. But the skeptic cannot claim victory either.

Laxminarayan Lenka's paper titled "Skepticism and Dretske's Zebra" attempts to look at one important assumption that the skeptic makes. This is the principle of knowledge being closed under deduction, also known as the Principle of Knowledge Closure (PKC). The paper first highlights that the PKC is employed in the Dream Argument, the Evil Demon Argument, and the Brain-in-the-Vat Argument and explicates how the PKC is employed to draw out skeptical consequences concerning knowledge about the external world. Lenka then briefly considers F. Dretske's Zebra example as an attempt to challenge the PKC. The discussion is followed by criticisms, against the Zebra example, by P.D. Klein as well as G. Stine. Lenka then offers an argument to show that the criticisms made by Klein as well as Stine may be off the mark and the Dretskian argument can be defended especially in cases that involve knowledge of logical and mathematical truths. An important question, however, remains. Will the defence work for knowledge claims regarding external world? After all, Dretske's Zebra example is an attempt to challenge the application of the PKC in cases regarding the possibility of having knowledge about the external world.

C.A. Tomy in his paper, titled "Naiyāyika Externalism and Gettier-like Counter Examples," attempts to show that the Nyāya theory of knowledge represents a synthesis of what is in the Western epistemological tradition known as the internalist and the externalist accounts. Tomy first sets out the conditions that need to be satisfied, according to Nyāya, for any cognition to be a *pramā*. One of these conditions relates to the justification for the truth of awareness which is a cognizing event. The paper then proceeds to argue for the Naiyāyika version of externalism. The author first posits some of the internalists' views, e.g. those of the Mīmāṁsakas. Next, he employs B.K. Matilal's paraphrasing of Naiyāyika externalism both to bring out the difference between the internalists' and the externalists' positions in the Indian philosophical tradition and to highlight some of the objections to the Naiyāyika position especially the counter examples proposed by Sriharṣa. The paper then considers Matilal's attempt to develop a Nyāya response, in terms of the distinction between knowledge simpliciter and knowledge of the knowledgehood of the knowledge simpliciter, to these counter examples and argues that the attempt is unsuccessful. Next, Tomy attempts to employ A. Goldman's strategy of causal account of justification and claims that Goldman's analysis can satisfactorily handle the counter examples posed and some other Gettier-like counter examples raised by G. Harman. Tomy suggests that knowledge must be treated relationally and knowledgehood is a property belonging to a group of awareness episodes related among themselves in a certain way and with the world outside causally. It is worth considering whether this account of knowledge does justice to the Naiyāyika view and also adequately handles the counter examples posed by Śrīharṣa and Gettier-type considerations.

K. Srinivas, in his paper titled "B.K. Matilal's Analysis of Indian Approaches to Epistemology," suggests that the Western method of analysis is inadequate to illuminate the "structural depth" of classical Indian philosophical theories. Srinivas sets out, following Matilal, the nature of epistemology as practised by philosophers in both the Indian and the Western traditions. In elaborating further, the author recognizes that the empiricist tradition in the West emphasizes sense-experience as a source of immediate knowledge and seeks to show how that emphasis is prima facie incompatible with empiricism. He also suggests, paraphrasing Matilal, that there are further problems with the empiricism of both the Indian and the Western varieties of epistemology. Knowledge is "logically" constructed out of experience. However, these experiences as edifice of knowledge are appearances. In support of objective knowledge, the *pramāṇa* theorists attempt to bridge the subjective and the objective elements through the requirement of perceptual basis. According to Srinivas, Matilal interprets the *pramāṇa* theorists to subscribe to a form of weak empiricism in that sense experience is a starting point for knowledge.

Nirmalya Narayan Chakraborty, in his paper, titled "Rationality, Thought and Language," argues that ascription of rationality is ascription of propositional attitudes like belief. To ascribe belief is to ascribe holding of some sentences as true. Thus, possession of language and the theory of interpretation that is associated with a language is a necessary condition of rationality. Since language is a social act, rationality is a property of the community where members interact and sharing a world constitutes a necessary condition of such an interaction. So also is the interaction of the community with that world and this interaction invariably is a historical phenomenon. To say that machines think is to overlook all these seminal points about thought.

Smita Sirker, in her paper, titled "Knowledge, Belief and Human Rationality," sets out to study the connection between our beliefs and reasoning. She begins with a background of Cartesian foundationalism that underscores the justified true belief account of knowledge. The paper, in this context, also sketches the formation of beliefs, revision of beliefs, and impact of these beliefs on reasoning. Sirker then goes on to look at critiques of foundationalism and discusses in some detail the positions of various empiricist and quasi-empiricist theories of knowledge posed by W.V. Quine and A. Goldman. The paper then engages with the question of human rationality and suggests that such rationality involves not merely drawing correct or incorrect inferences but also cognitive, psychological processes. The author while summarizing the argument from charity, the argument from evolution, the argument from reflexive equilibrium, and the argument from content, takes the stand that human reasoning is amenable to both descriptive and normative models without there being any systematic gap between the models. The actual observed performance in reasoning is not strikingly different from the normative dictates. The author also points out that humans possess unjustified and even false beliefs. And in our reasoning our entire belief corpus — true and false beliefs — are implicated. Sirker also argues that beliefs manoeuvre and direct reasoning. Beliefs have an important role in our decision making and actions.

Bipin Indurkhya's paper, titled "Rationality and Reasoning with Metaphors," deals with the seminal role of metaphors in reasoning and cognition. The traditional accounts of metaphors were dominated by the view that metaphors were at best pedagogical devices and at worst mere embellishments that can be reduced to the literal descriptions without any cognitive loss. Given the pervasiveness of metaphors in our cognitive practices, the question arises, "when is it rational to accept a conclusion derived from a metaphor?" Indurkhya deals extensively with this question. An adequate answer to this question, according to him, will result in the reciprocal effect in the sense it alters considerably both our conception of rationality and our own idea of thinking through metaphors. In working out his answer, Indurkhya draws our attention to two modes of metaphorical reasoning — one which is similarity-based which he calls "monotonic" and the other non-similarity-based which he calls "non-monotonic." For the reasons he convincingly articulates he attaches greater significance to the latter whose texture he elaborates upon very effectively. Taking examples from product-development, jurisprudence, and history of science, Indurkhya shows how non-montonic reasoning operates with metaphors that create similarities (rather than basing on existing ones) between the source-phenomenon and the target-phenomenon. Particularly significant in this connection is his claim that the creative metaphors alter the ontology of the target-phenomenon thereby creating new perspectives. In elucidating this point he invokes the view of cognition as an interaction between the cognitive agent, its representations of phenomena and the world itself which acts as the cognitive agent's environment. Indurkhya explains how the cognitive agent alters its concept-network to work out an ontology for the domain of phenomena that calls for a new mode of explanation in the face of what he calls "incoherency" (similar to what Kuhn calls "anomaly" or what he in his later work calls the violation of "no overlap" principle). Hence, the rationality of the cognitive agent consists, not in predictive success, but in the attitude of the cognitive agent that acknowledges "incoherency" and the ability to overcome it by making ontology-changing moves, thereby recognizing the autonomous structure of the world.

Amitabha Gupta, in his paper, titled, "Epistemology and Knowledge Representation: A Cognitive-Historical Approach to Philosophy of Science," first sets out the basic principles of the standard epistemological view. He characterizes this view as a result of the Logical Postivistic and Post-Postivistic articulation of the epistemological norms. This standard epistemological view, according to Gupta, characterizes the nature of scientific knowledge and some of its highly acclaimed features like truth, objectivity and explanatory efficacy in terms of the syntactic and propositional approach to scientific knowledge representation and revision. This approach offers an account of how scientific beliefs are formed or scientific laws, theories or models are constructed and revised. The author then summarizes some of the well-known reasons for dissatisfaction with this approach. Subsequently, Gupta, following the literature, explores a new set of reasons for further exposing the weaknesses of the syntactic and propositional approach. These new reasons arise from the more recent results in cognitive science which seem to show that the standard approach obscures a number of cognitively and epistemologically significant features of scientific knowledge representation and revision. The paper then attempts to make a case for sub-symbolic approach to scientific knowledge representation and revision on cognitive-historical and epistemological lines.

In Part II, there are four papers which are related to the general theme of Cognition and Science. Chhanda Gupta, in her paper, titled "Explanation, Causation and Physical Necessity: From the Standpoint of Internal Realism," seeks to show that we can retain against empiricists the notion of physical necessity that figures in scientific explanations even while demystifying it by eliminating dubious references to theological and metaphysical notions. Central to her attempt is Internal Realism championed by, among others, H. Putnam (especially in his later writings). According to the Internal realists we only have adequate empirical discourse about what there is if whatever we are talking about is relatable to our thoughts, concepts, and experiences. The stuff of reality has a composition which our theory conceives it to have. The Internal Realists seek to explicate "physical necessity" in terms of "rigid designation," "rigid designation" in terms of "substance-identity" and endeavour to construe "substance-identity" in a theory-relative way. The author juxtaposes S. Kripke's notion of "metaphysical necessity," which is radically non-epistemic, with Putnam's notion of "physical necessity" which is moderately non-epistemic in the sense it is something which is not just a projection. To obviate the skeptical challenge arising from radical changes that scientific theories undergo, the Internal realists take recourse to what is called "Transcendental Argument." According to this argument, the very fact that communication is possible between followers of any two inconsistent theories shows that there is coherence in their beliefs. The possibility of coherence in its turn shows that at least some of the beliefs are true.

Kausik Bhattacharya, in his paper, titled, "The Observational-Theoretical Distinction Revisited," argues that the philosophers who question observation-theory distinction base their position on the claim that observational terms stand for dispositions/causal properties and hence entail laws/generalizations which thus negate their purely observational characteristics. Bhattacharya seeks to show how such a claim leads to the anti-distinction view of K.R. Popper to conflict with the falsification criterion of Popper. Quine maintains that our statements about external world face the tribunal of experience not individually but as a corporate body. Yet Quine allows re-evaluation of any particular statement in the face of recalcitrant experience. But such a re-evaluation demands that observational terms stand for things that are directly observable and not their dispositional or causal properties. Hence, the rejection of observational-theoretical dichotomy is highly problematic.

Milan Desai, in her paper, titled, "On Inter-theoretic Incommensurability in Science," contends that Popper's semantic view is completely at odds with his falsificationist theory. Acceptance of the view that universal terms are theory-laden lands us into incommensurability and the consequent relativism. After considering the theories of meaning put forth by Kripke and Putnam as inadequate and conventionalism of Quine, she sees a way out of the quagmire of relativism in the *sphoṭa* theory of meaning which is neither intensionalist nor extensionalist nor conventionalist.

Manjari Chakrabarty, in her paper, titled, "Creativity and the Growth of Scientific Knowledge," questions the invidious distinction between the context of discovery and the context of justification made by Popper and his claim that the former falls outside the purview of Philosophy of Science as a Rational Reconstruction of Science. She delineates the central features of the discovery process as phenomena amenable to logical articulation that can shed new light on the nature of creativity. The author brings to bear on the problem of creativity the most recent work of philosophers, cognitive scientists, and historians of science in particular regarding the ways in which a creative breakthrough relates itself to and yet departs from the cognitive context from which it emanates.

In Part III of this volume, the three essays attempt to address three different issues that arise in the context of Cognition and Mind. The paper, by Amita Chatterjee, titled "How did the Faculty of Language Evolve?: A Debate Between Chomsky and Pinker," is an attempt to critique the position of M.D. Hauser, N. Chomsky and W.T. Fitch, which puts forth a version of "Faculty of Language" thesis that seeks to combine the linguisitic construal of language in terms of an abstract core of computational operations and a biological and psychological view of language as a uniquely human communication system. Hauser, Chomsky and Fitch seek to bring the "faculty of language" thesis close to evidence by adopting a bio-linguistic perspective. However, the theory is at odds with Chomsky's central contention about language. That apart, their identification of the evolutionary dimension of the faculty of language is based on a unilateral relationship between niche and the species. This has been called into question by cognitive scientists (and also evolutionists like R. Lewontin).

Manidipa Sen, in her paper, titled "Content Externalism versus Eliminative Materialism: A Case Against Scientism," argues that the eliminative theory of mind is inadequate by supporting G. McCulloch's attack on eliminativism which is ably articulated by P. Churchland. Churchland's scientific eliminative materialism is, according to McCulloch, guilty of incoherence. McCulloch's theory of mind, even while rejecting eliminativism, steers clear of Cartesian dualism. While sharing externalism with D. Davidson, he works out a radical version of it which is called content externalism. According to Sen, such a view, though rejects the ontological distinction between the subjective and the objective, nevertheless can maintain an epistemological distinction between the ways we come to know the mind and the ways we come to know the physical world — a distinction which further distances the view from externalism.

Madhucchanda Sen, in her paper, titled, "Empirical Knowledge and the Inner-Outer Distinction," tries to show how the hold of the dichotomies like inner-outer can be loosened by drawing our attention to philosophically rich insights of J. McDowell who very ably rejects the notion of the given — the notion central to all the attempts to ward off the challenge of skepticism. McDowell uses the Kantian notion of receptivity and spontaneity to work out his theory of experience as conceptual. Sen ably discusses the non-conceptualist criticism against McDowell and defends the latter. She brings out lucidly the distinction between the views of D. Davidson and McDowell both of whom reject the notion of Given and thus foundationalism. Particularly significant is her identification of the main thrust of McDowell's anti-Cartesian philosophy of mind.

PART I

Epistemology and Cognition

1

Six Responses to Skepticism

Sandhya Basu

THERE has been no dearth of skeptics and no dearth of anti-skeptics or epistemologists in the history of philosophy, ancient or modern, East or West.

The present paper confines itself to a brief exposition of the skeptical argument/arguments and a few responses to the skeptical argument/move offered by contemporary analytic philosophers. What I have written in this paper has no claim to originality. Each and every idea utilized here I have borrowed knowingly/unknowingly from some great philosophers or not so great philosophers and my lived experience. We all know that philosophical skepticism goes back to antiquity. Pyrroh of Elis and his followers were skeptics of ancient Greece. In Sextus Empiricus's *Outlines of Pyrrohnism*[1] we find an excellent exposition of the best developed form of skepticism where we get a very clear picture of the ancient tradition.

I

But the first question is: What is philosophical skepticism? In saying that the philosophers' doubts are neither confirmed

1. Sextus Empiricus, *Outlines of Pyrrhonism*, English tr. Rev. R.G. Bury, Cambridge, Mass: Harvard University Press, 1976.

nor confuted by empirical evidence Ayer identifies what philosophical skepticism is not. In my search for a positive reply I can begin by the remark that philosophical skepticism is questioning the truth of a belief (philosophical/popular) about the possibility of human knowledge. To put flesh and blood into this skeleton remark let us take help from Ayer. Ayer observes that the skeptic's "argument is designed to show that however circumspect we are it makes no difference. . . ."[2] That is why the skeptic never encourages us to be more circumspect. He casts doubt upon the possibility of knowledge as such. The skeptic doubts whether under any circumstances existence of physical objects can be proved by sense-perception; whether any inference from past to future or from present to past is legitimate; whether memory can at all be trusted; and whether we can ever have knowledge of other minds.

> The skeptic is, strictly, not one who denies the validity of certain types of belief, but one who questions, if only initially and for methodological reasons, the adequacy of our grounds for holding them. He puts forward his doubts by way of challenge — sometimes a challenge to himself — to show that the doubts are unjustified, that the beliefs put in question are justified.[3]

The fact that philosophical skepticism cannot be confuted by experience, as is claimed by the skeptics, makes it too ambitious to freely fly to any height without having the fear of being contradicted by experience. Following in the footsteps of great philosophers I hope to show that this tall claim is vacuous and the flight is just imaginary.

---

2. A.J. Ayer, *The Problem of Knowledge,* Middlesex: Penguin Books Ltd., 1956, p. 37.
3. P.F. Strawson, *Skepticism & Naturalism, Some Varieties,* Great Britain: Methuen & Co Ltd., 1987, p. 2.

Another important point is that skepticism is necessarily epistemological in nature. To anyone's claim that there is metaphysical skepticism as well, my reaction would be: even then epistemological skepticism is logically prior to metaphysical skepticism. The structure of the skeptical argument is like this: Knowledge is possible only if a certain set of conditions is fulfilled. It is impossible to fulfil all the conditions. Therefore, it is impossible to have knowledge. This skepticism might be global or local in nature. Global skepticism is the view that we know nothing or that everything is open to doubt. The global skeptic thus says that we know that we do not know anything. This sounds obviously paradoxical. We will be back to this point very soon. The local skeptic, however, takes up the accredited sources of knowledge and on scrutiny abandons some. And depending on this, or even independently, he may be skeptical about the claims of moral philosophers, or about religious beliefs, or about knowledge of other minds, about time or about knowledge, about the past or future. In a different way of putting the matter, he may doubt perception or inference or testimony or *anupalabdhi* or intuition or some other source as source of knowledge at all.

Now, a return to global skepticism. Elaborating on what we have said in the last paragraph, it can be said that knowledge requires indefeasibly justified true belief. But with regard to no belief can we be sure that it is indefeasible. So we can never be sure that we have knowledge. Again, there are some skeptics who think that not to speak of having knowledge, we do not know what knowledge is. For we cannot identify the instances of knowledge. In order to identify the instances of knowledge we need some definition of knowledge. And, in order to define knowledge we have to discriminate the samples of knowledge from samples of non-knowledge. The skeptic's answer to this famous problem of criterion is that knowledge is impossible. Since "it rains alike on the just and the unjust"[4] it is supposed to be easily refutable. We have no means of discriminating between the just and the unjust, between the infallible and the fallible, hence we can't help submitting ourselves to an all-inclusive fallibilism. Both Ayer and Ryle think of a rather simple retort at first. Let's hear from them one after another. "A perception is called illusory by contrast with other perceptions which are veridical; therefore to maintain that all perceptions must be illusory would be to deprive the word 'illusory' of its meaning."[5] Hence there must be some veridical perceptions. Ayer came to realize pretty soon that his argument is indecisive. His rejoinder follows: "Even granting that it makes no sense to say that all our perceptions are delusive, any one of them still may be. We have to make good our claim to know that some particular ones are not."[6] This very line of reasoning both for presenting the skeptical position and going against it, Ryle had recourse to. "There can be false coins only when there are coins made of the proper materials by the proper authorities."[7] I quote Ryle again:

> . . . however general his (a layman's) suspicions might be, there remains one proposition which he cannot entertain, the proposition, namely, that it is possible that all coins are counterfeit. For there must be an answer to the question counterfeits of what?[8, 9]

The non-decisive character of this argumentation is at par

4. A.J. Ayer, op. cit., p. 38.
5. Ibid., p. 37.
6. Ibid., p. 38.
7. G. Ryle, *Dilemmas*, Cambridge: Cambridge University Press, 1977, p. 95.
8. Ibid., pp. 94-95.
9. Just as we cannot talk of appearances without admitting something which appears.

with that of Ayer's argument. This reminds us of Chisholm's anti-Pyrrhonian principle, "If anything is probable for S, then something is certain for S."[10, 11] James Cleave,[12] however, has something to say to people who are not convinced by Ayer's or Chisholm's principle (which was used in a different context though). It is this: in order to argue that some perceptions are non-veridical or that some coins are counterfeit, or that something is probable it is enough to *assume* that some other perceptions are veridical, or that some coins are genuine, or that something is certain. We need not *prove* the veridicality of perceptions, genuineness of coins, or certainty of propositions. Perceptions which are supposed to be veridical, or coins considered as genuine, or propositions taken to be certain need not actually be so. We shall come back to this position and its influence on skepticism later in a more appropriate context. Ayer comments:

> From the fact that our rejection of some of them (judgement) is grounded on our acceptance of others it does not follow that those that we accept are true.[13]

II

According to Thomas Nagel skepticism is the inevitable outcome of any sustained attempt to command an objective view of one's relations to the world. There are many cynical

10. R.M. Chisholm, *Theory of Knowledge*, 3rd edn. New Delhi: Prentice Hall of India Private Limited, 1992, p. 14.
11. Compare L. Wittgenstein, *On Certainty*, G.E.M. Anscombe & G.H. von Wright (eds.), D. Paul and G.E.M. Anscombe, New York: Harper Torchbooke, Harper & Row Publisher, 1972. Proposition 115: "The game of doubting itself presupposes certainty."
12. J. Cleve, "Foundationalism, Epistemic Principles, and the Cartesian Circle", *The Philosophical Review*, LXXXVIII, no. 1. 1979.
13. A.J. Ayer, op. cit., p. 38.

non-skeptics who contend that the attempt to meet or even to understand the skeptical challenge is an idle exercise, a wilful refusal to abandon outmoded forms of thinking.

Peter Unger tries to defend skepticism in the following way. He holds that the attitude of absolute certainty (which he equates with knowledge) is never all right for anyone ever to have, provided there are no overriding considerations. Hence we can come to the conclusion that nobody ever knows anything about anything. "No matter how comfortable we may feel in our philosophy and our view of the world," says Unger, "I cannot see how we might properly be *certain* that there is *no way* that we could possibly be wrong."[14] In another paper, "I do not exist,"[15] Unger takes "Sorites Decomposition" argument (a simple form of Sorites Paradox) as his main argument for proving that the skeptical position is justified. He applies the argument even to the existence of himself, the existence of "I." If I exist I consist of a finite number of cells. Removal of one or few cells will not make any difference to me. If, thus, in a finite number of moves all of my cells are removed I will still be there. But such an existence is indistinguishable from non-existence. So initially, I do not exist, I never have or never shall. This argument can be refuted either by challenging premises (1) I exist; (2) I consist of a finite number of cells, or by challenging the argument itself.

III

Local skepticism is more thoroughly worked out in the Indian

---

14. P. Unger, *Ignorance: A Case for Skepticism,* Oxford: Clarendon Press, 1975, p. 134.

15. P. Unger, "I do not exist", in G.F. Macdonald (ed.), *Perception and Identity,* London: Macmillan Press Ltd., 1981.

tradition than in the Western tradition.[16] In *Tattvopaplavasiṁha* Jayarāśi attacks every claim to knowledge. Testimonial knowledge is a very easy target. Many people think that scientific training advises us not to take anything on trust, on testimony. Unlike these there are perhaps many more who think that scientific progress of a society is measured by how much is accepted on testimony. The more advanced a society is the more it depends on books and journals and newspapers and television news and all other recent informative technological devices.

How is testimony attacked? A follower of Nyāya theory says that testimony is reliable because it is uttered by infallible persons. But how do we know that the persons concerned are infallible or *āpta*? Again, *Tattvopaplavasiṁha* asks: Is being *āpta* enough or do we have to know their *āptatva* in order to have testimonial knowledge? Then the author of the said book attacks *apauruṣeyatva*. According to the Mīmāṁsakas, the uncaused, unauthored is called *apauruṣeya*. Suppose we find the following *apauruṣeya naisargika* statement. There is a stone in a river from time immemorial. Just by natural course there are inscribed on the stone the words "Vedas are false." Since these words are not authored by anybody they are *apauruṣeya* and consequently true. Is it not absurd? Again, testimony is reduced to inference and then criticized by use of standard arguments against inference. Third, it is difficult to decide which scriptures are true and which false, for there are conflicts among scriptures themselves.

Let us next discuss skeptical attacks against perception. *Tattvopaplavasiṁha* is the only book in Indian epistemology

16. This section has been written solely on the basis of notes of lectures in a Refresher course delivered by Professor Arindam Chakrabarti in Lucknow in 1988.

which attacks and refutes *pratyakṣa*. It finds hundreds of holes in *pratyakṣa*. All definitions of perception, it argues, are wrong. And if one fails to define then one cannot claim to have perceptual knowledge. Next, reliability of perception is questioned. One stock argument is: perception is known to be veridical if it is known that it is not due to a defective sense organ. But how can we know the non-defective character of a sense organ? Surely sense-perception cannot help us here for an *indriya* is *atīndriya*. Neither does coherence help us here. For there are many dream-experiences which are fully coherent. Thus, definability and reliability of perception are both refuted.

IV

Consideration of the responses to skepticism can very well begin with the response given by Moore who in Peter Unger's opinion is "perhaps the most influential opponent of skepticism in this century." But before that it's not a bad idea to go back to David Hume whose "philosophy," in Mohanty's words, "is an attempt to assign man's natural beliefs and sentiments their rightful place and their legitimate status in spite of the fact that reason is at variance with them."[17] Hume's writings perfectly conform to the above picture given by Mohanty. I quote: "The great subverter of Pyrrhonism or the excessive principles of skepticism is action, an employment and the occupations of common life."[18] In continuation, in the next page, Hume comes up very boldly and very sharply, " . . . though a Pyrrhonian may throw himself or others into a momentary amazement and confusion by his profound reasonings, the first and most trivial event in life will put to flight all his doubts

17. D. Hume, *An Enquiry Concerning Human Understanding*, Calcutta: Progressive Publishers, 1982; See the "Introduction", p. xiii.

18. Ibid., p. 126.

and scruples, and leave him the same, in every point of action and speculation. . . ."[19] For "Nature is always too strong for principle." And human nature is by nature disposed to beliefs in the external world and such other things which are disputed by the skeptic. This kind of naturalistic solution, however, is not to everyone's liking and is not given a top ranking. As we all know Moore seeks to repute the skeptical position by holding up his two hands while standing and making certain gesture with the right hand when he says "Here is one hand," and a certain gesture with the left hand when he says "And here is another." Moore thinks that he is thus offering a "perfectly rigorous" proof of the existence of external things and that no more rigorous proof can be had. Moore, the particularist, thus comes up with boldly claiming that we can never be more certain about the truth of any general principle than about the truth of any particular propositions like "This is a hand." Possibility of dreaming may seem to refute Moore's claim to knowledge. If Moore does not know that he is not dreaming then he does not know that he is standing up. If it is true, it follows that it is also true that if he does know that he is standing up, then he does know that he is not dreaming. Moore therefore argues: since he does know that he is standing up, he does know that he is not dreaming. The opponent's argument comes immediately: since Moore does not know that he is not dreaming, it follows as a consequence that he does not know that he is standing up. Moore holds that the *one argument is just as good or as bad as the other.* If the opponent cannot give a stronger and thus better argument then Moore has every right not to give up his empiricist particularist position. Moore thinks that he has conclusive reasons for asserting that he is not dreaming, "conclusive evidence" that he is wide awake.

19. D. Hume, op. cit., p. 127.

Some philosophers, including Ludwig Wittgenstein who was full of admiration of Moore's honesty, depth and seriousness, are of the opinion that Moore's argument is just not relevant to philosophical skepticism. Norman Malcolm in his book, *Ludwig Wittgenstein: A Memoir,*[20] gives a beautiful picture of Wittgenstein's comments on Moore's refutation. I quote Malcolm:

> "Moore would like to stare at a house that is only 20 feet away and say with a peculiar intonation, *"I know that there's a house!"* He does this because he wants to produce in himself the *feeling* of knowing. He wants to exhibit *knowing for certain* to himself. In this way he has the idea that he is replying to the skeptical philosopher who claims that everyday examples of knowing . . . are not really or strictly *knowledge*. . . . It is as if someone had said "you don't really feel pain when you are pinched" and Moore then pinched himself in order to feel the pain, and thus prove to himself that the other is wrong.[21]

Moore does not hesitate to add that when he asserts that he is standing up he is aware of the fact that he might have been sitting down at that very moment, and in that case his assertion would have been false. But there is a very important difference between "might have been sitting down" and "actually sitting down," and between "would have been false" and "is false." The proposition "I am standing up" is not false, it is true as a matter of fact, though not as a matter of necessity.

Barry Stroud categorically says that Moore has not been successful in his attempt, for skeptics bring everything into question at one fellswoop. Internal reactions or autobiographical statements are not sufficient for rebuttal of skepticism

20. N. Malcolm, *Ludwig Wittgenstein: A Memoir*, Oxford: Oxford University Press, 1984.

21. Ibid., p. 71.

which is in need of an external answer. Even Strawson goes against Moore when he says that either Moore misses the point of the skeptical challenge or has recourse to an unacceptable dogma.

V

A "brilliant novelty of Wittgenstein's thinking" (Malcolm) is illustrated in his refutation of skepticism as found in his *On Certainty*[22] and *Philosophical Investigations.*[23]

Wittgenstein argues that philosophers and non-philosophers alike have to participate in the language game. Most games are meant for recreation. But the game of language is different. Language-game is the most important mode of communication. Like most other games the game of language is also rule-governed. But in many cases the native speakers of the language frame the rules as they go on using the language and there are occasions which demand alteration or revision or moderation of the rules, without abandoning them *en bloc.* And one cannot participate in this game unless some propositions are assumed to be beyond question and doubt. These are the propositions which belong to our "frame of reference." They are called framework propositions doubting or questioning which entails that the language game cannot get started.

Let us try to string together some of Wittgenstein's main remarks scattered here and there in his writings. The proposition "I know that p" is meaningful only in the context of language-game. Words have their meaning only in the flow of life.[24] As propositions which are not open to doubt and not

22. L. Wittgenstein, *On Certainty,* op. cit.

23. L. Wittgenstein, *Philsoophical Investigations,* tr. G.E.M. Anscombe, Oxford: Basil Blackwell, 1984.

24. L. Wittgenstein, *On Certainty,* op. cit., p. 84.

open to empirical test, the framework propositions are assumed to be true, but not absolute in the sense of beyond change. The language-game also changes generally over time. And this is accompanied or followed by change in concepts and consequently the meanings or words change. Comparing our thinking with water flowing down the river Wittgenstein says in proposition 99:

> . . . the bank of river consists partly of hard rock, subject to no alternation or only to an imperceptible one, partly of sand, which now in one place now in another gets washed away or deposited.[25]

> While the form or structure of a phenomenon continues to appear perceptually identical with or the same as its "inner" stuff, matter or constituent keeps on changing, decaying, transforming, etc. Human bodies illustrate the point clearly.[26]

As "the scaffolding of our thoughts" the framework propositions are, Wittgenstein says, exempt from doubt. Let's have a look at propositions 343 and 344 of *On Certainty*. Part of 343 says: we just *can't* investigate everything, and for that reason we are forced to rest content with assumption. If I want the door to turn, the hinges must stay put. Proposition 344: My *life* consists in my being content to accept many things.

Wittgenstein is critical about Moore's method of refutation of skepticism and the examples cited by the latter are ridiculed by Wittgenstein as lacking relevance and significance. When the skeptics say "you don't know" and Moore replies "I do know" his reply is just useless and beside the point. No one,

---

25. L. Wittgenstein, *On Certainty*, op. cit., p. 84.
26. D.P. Chattopadhyaya, "On Some Ways of Getting Rid of Scepticism", *Realism: Repsonses and Reactions*, eds. D.P. Chattopadhyaya et al., New Delhi: ICPR, 2000, pp. 311-12.

neither the skeptic, nor Wittgenstein nor Malcolm doubts that we have feelings of certainty about ordinary propositions like "This is a tree," "This is my hand," "The earth has existed for many years." Autobiographical reports expressing one's own feelings and emotions are not what the skeptic is interested in. Personal attitudes may find an expression in these examples — but they cannot instantiate knowledge. Transmissibility of knowledge from the knowing subject to others is sufficient to prove objectivity of knowledge. One's claim to knowledge must be legitimized or validated by general consent of the community. This is "Social Naturalism." When everyone around me can see the tree or the hand isn't it crazy to utter sentences like "This is a hand," "This is a tree?" These are like saying "Good morning!" or "Congratulation!" in the middle of a conversation about, say, artificial intelligence or paucity of brilliant students in the departments of philosophy across the globe. Patients suffering from some degree of Alzheimer's sometimes come out with uttering irrelevant sentences.

In my opinion Wittgenstein in not being just to Moore. According to Moore, the examples he has given are perfectly relevant. If we cannot claim knowledge in these simple cases then we can keep mum. Moore wants to begin with cases where the skeptic is unable to throw his arrow. Moore is never worried about his subjective conviction or personal feelings. He is fully aware of the fact that the skeptic's target is knowledge and I think, Moore's robust ordinary language philosophy is not a misfire.

In concluding this section I am being more and more inclined to the view that Wittgenstein is not really opposed to Moore's view. Let us quote proposition no. 7 of *On Certainty:*

> My life shows that I know or am certain that there is a chair over there, or a door, and so on — I tell a friend, e.g. "Take

> that chair over there," "Shut the door," etc. etc.[27] Again in proposition 10 he says: ". . . I know that there's a sick man lying here," used in an *unsuitable* occasion, (say, when I am at his bedside)[28] seems not to be nonsense but rather seems matter-of-course, only because one can fairly easily imagine a situation to fit it, and one thinks that the words "I know that . . ." are always in place where there is no doubt . . .[29]

In fact, what Wittgenstein opposes in Moore's view is the latter's use of examples in some inappropriate situations, not that all his examples are out of place or misfit. And there is no reason why the idea of "language game," which is just a form of life would unnerve Moore. The phrase may be due to Wittgenstein, but we the so-called educated persons all know that we participate in the game of language for the purpose of communication and it is we, the users of language, who frame rules for the use of expressions.[30] It has appeared to me that about rules of the use of the word "Knowledge" there is hardly any difference between Moore and Wittgenstein.[31] And if both of them can show, may be in not exactly the same way, that knowledge is possible we can live in a safe distance from the skeptic.

VI

R.M. Chisholm in his monograph, *The Problem of the Criterion*,[32] offers a very simple answer to the skeptical challenge. A strong

---

27. L. Wittgenstein, *On Certainty*, op. cit., p. 3e.
28. Ibid. Words within brackets are my addition.
29. Ibid., p. 3e.
30. P.F. Strawson, *Introduction to Logical Theory*, Oxford: Methuen & Co. Ltd., 1971.
31. Some argue that Moore did not influence Wittgenstein at all.
32. R.M. Chisholm, *The Problem of the Criterion*, Milwaukee: Marquette University Press, 1973.

advocate of St. Augustine's philosophy Chisholm observes: ". . . The senses should be regarded as innocent until there is some positive reason on some particular occasion for thinking that they are guilty on that particular occasion."[33] While in search of a solution to the problem of the criterion he chooses the Moorean form of particularism and says that we are more justified in believing particular propositions than believing general philosophical principles, though Moore's examples he cancels as paradigms of absolute certainty. Of course, one may be skeptical about the status of "Particular propositions are more justified than general principle." Does Chisholm have to handle a paradox here? Perhaps he can avoid it by saying that if it is paradoxical, then every particularist has to face it. Or, he can say that some presuppositions and some solutions must be left out as not falling within the fiery raining zone of the skeptic. The present author is in favour of the latter alternative.

P.F. Strawson offers a realist-cum-naturalist response to skepticism, and the source of inspiration is Humean naturalism. We cannot help accepting belief in the existence of body, in the reliability of induction, in general veridicality of sensory reports, for abandonment of such beliefs takes away our capability of rational questioning most of our beliefs. Hence reasoned rebuttal of skepticism is a futile academic exercise. To quote Strawson,

> To attempt to confront the professional skeptical doubt with arguments in support of these beliefs, with rational justifications, is simply to show a total misunderstanding of the role they actually play in our belief-system.[34]

---

33. R.M. Chisholm, op. cit., p. 24.

34. P.F. Strawson, *Skepticism and Naturalism: Some Varieties*, London: (University Paperback) Methuen & Co. Ltd., 1987.

VII

Naturalistic response to skepticism is also found in Indian philosophical thinking. In spite of all our arguments against skepticism we have to have ordinary life. We have to define, we have to take help of *pramāṇa*, we have to claim knowledge. The only difference between a skeptic and an ordinary man is that the skeptic's beliefs are *admittedly* irrational, whereas the ordinary man's irrational beliefs are not recognized by him to be irrational.

Against the skeptic Udayanācārya offers a typical realistic argument in his *Ātmatattvaviveka*. He compares the skeptical move with a fictional story. On hearing the story of the *Rāmāyaṇa* a young monkey tries to cross the ocean and fails and then comments: *apāroyam akupāraḥ pārāvāraḥ mithyā rāmāyaṇa kathā*. Since the monkey fails in an endeavour which was stated to be successful in *Rāmāyaṇa* in his eyes *Rāmāyaṇa* is false. This is unjust. The skeptic cannot refute the existence of the external word and our knowledge of the same in such a ridiculous way.[35]

VIII

The project of wholesale validation, observes Strawson, is a chimera. A dissection of our conceptual scheme for discovering and investigating the connections and interconnections between the major structural constituents, if successfully performed, can abate the skeptic. Who speaks of conceptual scheme and the major structural constituents? It is Kant and Strawson, the professed Kantian. Kant's categorial framework has no history, it is invariable and inter-subjective. D.P. Chattopadhyaya explicitly expresses his unhappiness about the position explained above. He writes,

---

35. See note 16.

> It leaves the relation between the "massive central core" area and the non-core area of changing commonsense and developing science ill-defined and incoherent. . . . The prohistoric critic is clearly entitled to extract a skeptical element from the incoherent account of the relation between the metaphysical core area and the empirical non-core or peripheral area.[36]

A person in love with Kant's philosophy is naturally blind to its pitfalls. But the philosopher in that person must wake up and be out of love for at least the time he is engaged in philosophical discourse. The charge of ahistoricity is difficult for him either to swallow or to shake off. The Kant of his *Critiques* has been reviewed and rewritten. No need of mourning and writing an elegy. No "Farewell to Kant" banner. A philosopher who has not read Kant is no philosopher — no one is willing to deny this. But we have to keep pace with progress just as we have to look back to the past, and this is history which we cannot refuse to efface.

IX[37]

In his paper, "A Coherence Theory of Truth and Knowledge,"[38] Donald Davidson attacks the global skeptic, transcendentally, not frontally.[39] Breaking from tradition Davidson contends that his coherence theory is about coherence of beliefs or of sentences held true by someone who understands them and

---

36. L. Wittgenstein, *On Certainty,* op. cit., pp. 320-21.

37. This section is an imperfect output of the input given by Professor Pranab Kumar Sen who initiated me into Davidson's philosophy.

38. D. Davidson, "A coherence Theory of Truth and Knowledge", in E. Lepore (ed.), *Truth and Interpretation: Perspectives on the Philosophy of Donald Davidson,* Oxford: Basil Blackwell Ltd., 1986.

39. See my paper, "A Long Essay on Donald Davidson's Coherence Theory", *JICPR,* New Delhi, vol. XIX, no. 3, 2002.

not about propositions. This coherence is constituted of logical consistency, inferential relations of various sorts, explanatory relations, relation of mutual support, etc. Our beliefs are identified and described only within a dense pattern of beliefs. We cannot entertain beliefs in isolation. We always hold beliefs that form a cluster. However, coherence by itself is no guarantor of truth. Coherence of beliefs is buttressed by the causal basis of beliefs, correct interpretations of other's speech, right ascriptions of beliefs and other propositional attitudes to other people. It is to be noted further that as long as we cannot reach a completely consistent set of beliefs all the beliefs in the set cannot be given equal weightage. Some must be acknowledged as more worthy than other beliefs. And coherence with these beliefs creates a presumption in favour of the truth of cohering beliefs. At this point Davidson and his followers cannot set aside the obvious fact that the goal of a completely consistent set is not reachable; at least in some discourses and disciplines it is an elusive goal. This coherence, however, is the test or criterion of truth, the nature of truth being explained by correspondence of beliefs with reality. Had our beliefs not corresponded with reality they could not cohere with one another.[40] Correspondence needs no confrontation with reality — direct handshaking with reality is a misleading idea. A deep metaphysical connection is there between beliefs'

40. Though transcendental arguments are very close to Kant's heart, about the truth of judgements his arguments are not perhaps strictly transcendental. What Kant says is that truth of judgements is proved by their communicability and universal validity. According to Kant successful communication and coherence of judgements of different individuals are the criteria of truth which is defined (unlike Davidson) by him as agreement of knowledge with its object (see N.K. Smith., (tr.), *Immanuel Kant's Critique of Pure Reason,* London: Macmillan & Co. Ltd., 1963, A820, A824 to B848).

correspondence and coherence. In a very important sense, believing is rooted in being. Effort for bridging the gulf between believing and being runs in the very blood of believing. This is why though any individual belief of ours might be false each and every belief of ours cannot be false. In proposition 232[41] Wittgenstein had said: "We could doubt every single one of these facts (*beliefs*) but we would not doubt them all."

Wouldn't it be more correct to say: "we do not doubt them all?"[42]

But what is the reason behind this attitude? The reason lies in making believers doubly sure against universal fallibility. What protects us from massive all-engulfing error? Davidson finds protection in the following argument. Suppose there is an interpreter who is omniscient, O_1, and thus, holds true beliefs and true beliefs alone. Now in order to interpret us, the human speakers, this O_1 must share many beliefs with us. And these shared beliefs must be true for they belong to O_1 also. If all or most of our beliefs were false then O_1 would fail to understand us and this is to deprive the omniscience of the O_1. Hence all our beliefs cannot be false — in fact, most of them are true.

However, truth of belief by itself does not amount to knowledge.[43] We need at least one more condition, viz. justification, to speak mildly, and indefeasible justification, to put it strongly. While holding true beliefs one may not have reasons to believe that beliefs are coherent and true. So

41. N. Malcolm, op. cit., Proposition 232.
42. See my book, *Justification: Concepts and Theories*, Kolkata: Progressive Publishers, 2002, Chapter Two.
43. Sandhya Basu, "A Long Essay on Donald Davidson's Coherence Theory," *JICPR*, New Delhi, vol. XIX, no. 3, 2002, p. 612.

justification may be absent. Again, it might happen that one has reasoned belief in the coherence of beliefs but does not appreciate one's reasons. Then, again, the question that is peeping is about indefeasibility of justification. It is not possible for me to answer all questions in the orbit of one paper. Let me just say one or two things here. Davidson's idea of "methodologically basic" beliefs and credibility of some occasional sentences having causal histories of their own may help us to some extent in getting closer to justification. I do not intend to confuse myself and the reader about the distinction between causal intermediaries and epistemic intermediaries. Now about indefeasibility — perhaps this is too tall an ambition. Insofar as empirical knowledge is concerned we can at most arrive at undefeated justification. The route of the journey has to be searched in Davidson's map of coherence theory.

Conclusion

The path that I have always chosen to walk along is an anti-skeptical one. Skepticism, I am inclined to think, is a way of life just as anti-skepticism is a way of life. We are free to choose either. But one may retort, "why have you written this paper then?" My reply may be — without consulting the specialists in the field I was not sure of my disease and its remedy. To satisfy the philosopher let us hear from a philosopher no less than Kant. I quote

> The skeptic is thus the taskmaster who constrains the dogmatic reasoner to develop a sound critique of the understanding and reason. When we have advanced thus far we need fear no further challenge, since we have learned to distinguish our real possessions from that which lies entirely outside them.[44]

---

44. Sandhya Basu, op. cit., p. 612.

2

Skepticism and Dretske's Zebra

Laxminarayan Lenka

THE first part of this paper contains an exposition of (a) The Dream Argument, (b) The Evil Genius Argument, and (c) The Brain-in-a-vat Argument. Each of these three arguments for skepticism presupposes the principle of Knowledge Closure (KC): If S knows P and knows that P entails Q, then, S knows that Q. In the second part of this paper, I make an exposition of Dretske's Zebra Case and criticisms made on it by Klein and Stine. In the third part, I propose the Tautology Case and, in the fourth part, I conclude by answering two questions having fundamental relevance to this paper.[1]

I

The Arguments for Skepticism

THE DREAM ARGUMENT

In *Meditations*, Descartes has advanced the Dream Argument

---

1. An earlier version of this paper was presented at a National Seminar on *Epistemology & Cognition in Comparative Perspective*, held at the Department of Philosophy, University of Hyderabad, India, in March 2007. I sincerely thank Prof. Suresh Chandra, C.A. Tomy, Madhuchhanda Sen, Rupa Bandopadhyaya, Manjari Chakrabarty, Venusha Tiny and Prof. Sandhya Basu for their active participation during the Open Discussion on my paper.

to deny the certainty of beliefs based on sense experience.[2] For Descartes, a denial of certainty is equated with the denial of indubitability, hence, the argument is meant to demonstrate how every belief based on sense experience can be subjected to doubt. Of course, "doubt" is methodological in the sense that he uses the method of doubting on every doubtable thing in order to arrive at the indubitable *qua* certainty (the *Cogito ergo sum*). But, instead of doubting the individual beliefs one after another, he doubted different sets of beliefs by doubting on the very principle or foundation on the basis of what a number of beliefs can be put into one set, group or category. The principle refuted by the Dream Argument is the principle of obtaining beliefs on the basis of sense experience.

Descartes has not stated the reason for which his way of grouping the beliefs (into two categories, namely, beliefs based on sense experience and beliefs which are very simple and have universal acceptance) is not subject to doubt. On the other hand, he gives a reason for which he makes the groupings. Beliefs can be infinite in number; hence, it is impossible to end the act of doubting if one goes on doubting beliefs, individually, one after another. Moreover, even if his classification is alright, one may very well doubt on the following argument:

A1:

P1: The Dream Argument is valid.

P2: If the Dream Argument is valid, then, nothing known on the basis of sense-experience is certain.

Therefore, nothing known on the basis of sense-experience is certain.

For the above argument is in *Modus Ponens* form and, as

2. E. Haldane, G.R.T. Ross, *Rene Descartes, Meditations on First Philosophy*, Cambridge Cambridge University Press, 1967.

we find later, even the principle of *Modus Ponens* is doubtable (*qua* not certain) on the basis of the Evil Genius Argument.

The Dream Argument can be presented as

A2:
P1: What we see in dreams need not be true.
P2: Our waking states are not distinguishable from our dreaming states.

Therefore, what we see even in our waking states need not be true.

That A2 is a valid argument makes the 1st premise of A1 true. The 2nd premise of A1 becomes true as the following two statements are true:

(i) In A2, any verb corresponding to a sense experience can replace "see" without affecting the validity of A2.

(ii) If something need not be true, it is not known. (What need not be true lacks certainty and, certainty being necessary for knowledge, what need not be true does not qualify to be called knowledge.)

Insofar as we are not certain that we are not dreaming now, we are not certain that what we experience now is there in reality and not in dreams. I see that I am in this room with others. The same might not be true but a dream; I might be sleeping in my bedroom and dreaming that I am in the class room. You may pinch me to make me realize that I am in the waking state and I may feel pain when you pinch me. But as the same can be a content of one of my dreams, what is the guarantee that I am not dreaming now? If everything I experience in a waking state can be experienced through dreams, no experience stands as the criterion to distinguish my waking state from my dreaming states.

This makes the 2nd premise of A2 true. As A2's 1st and

2nd premises are true, the validity of A2 makes A2 a sound argument. Of course, philosophers have challenged the soundness of A2 mainly on the ground that its 2nd premise is false. (Malcolm has strongly argued that the two states — waking and dream states — can be distinguished from each other.) We do not enter into the complexities of those arguments for or against the truth of A2's 2nd premise. My main concern here is to show how the principle of Knowledge Closure is presupposed by the Dream Argument.

In accordance to the Dream Argument, you do not know that p, if you do not know that q; when p stands for any proposition expressing an instance of your sense-experience and q for the proposition "You are not dreaming." You do not know that you are seeing me, even if you are seeing me, if you do not know that you are not dreaming. Of course, that you are actually seeing me and that you are dreaming that you are seeing me cannot be true together. But, why cannot you know the former if you are unable to distinguish it from the latter? The answer presupposes that the Principle of Knowledge Closure is true. The following argument is the answer.

A3:
If you know that p (= you are actually seeing me) and
You know that p entails q (you are not dreaming that you are seeing me), then,

You know that q.
In accordance to A3, since you do not know that q, either you do not know that p or you do not know that p entails q. As you know that p entails q, you do not know that p.

THE EVIL GENIUS ARGUMENT

Even if there is no criterion to distinguish the waking state from the dream state and, thereby, no criterion to ascertain that we are not dreaming now, this is not sufficient to establish that even the "simple" truths like 2+2=4, this object has a shape, an object is either red or not red and so on are doubtable. Knowing of such simple truths does not depend on the knowing of the fact that we are not dreaming now. Even if we are dreaming now, these truths cannot be false. Even in dreams, the simple truths do not fail. Such truths prevail in dream states as well as in waking states, hence, no distinction between the two states does not form a ground on which these truths can be doubted. This is the reason for which Descartes advanced his Evil Genius Argument over and above the Dream Argument.[3] Whereas the Dream Argument is alleged to be refuting the certainty of every possible belief based on sense-experience, the Evil Genius Argument is to refute the certainty of every possible simple truths of mathematics, geometry and logic or reasoning.

In accordance with the Evil Genius Argument, an evil genius (say, a demon) deceives all of us in such a way that, although the above simple truths are false, we mistake those false things to be simple universal truths. That is, for example, even if 2+2 is not equal to 4, we are deceived by the demon so that we believe that 2+2=4. The demon has full control over our minds and he makes us to believe that 2+2=4, he could have made us to believe that 2+2 is not equal to 4. The Evil Genius Argument can be presented as A4:

P1: If you know that p then you know that you are not deceived by the Evil Genius
P2: You do not know that you are not deceived by the Evil Genius
Therefore, you do not know that p.

Any criterion put forth for the knowing of not being

3. E. Haldane, op. cit.

deceived by the Evil Genius can be marked as an instance of being deceived by the Evil Genius. For example, the community consensus and universal acceptance of the rule of addition that makes 2+2=4 can be conceived of being so accepted or agreed upon due to the demon's control over the facts of the world as well as over our mind.

Whether there actually exists such an evil genius or not does not matter and the argument does not assert that it exists. What matters to the argument is the fact that we can conceive of such an evil genius and, thereby, it is logically possible that we are deceived by the Evil Genius. This possibility ensures that the so-called universal simple truths are without certainty and, thereby, as much doubtable as the truths based on sense experience.

Of course, it is logically possible that the demon deceives you. Also, that you do not know that it does not deceive you. But, then, why should not you know that p (say, 2+2=4)? That is, even if the 2nd premise of A4 is acceptable, how is its 1st premise acceptable? How does your knowledge of not being deceived by the Evil Genius necessarily follow from your knowing that p? The skeptic presupposes that the principle of KC is true and, in accordance to this principle, the following is a valid argument.

A5:
P1: You know that 2+2=4
P2: You know that if 2+2=4 then you are not deceived by the Evil Genius
Therefore, you know that you are not deceived by the Evil Genius
This is how, in accordance to the skeptical argument, it becomes necessary that you know that you are not deceived by the Evil Genius, if you know anything, say, you know that 2+2=4.

THE BRAIN IN A VAT ARGUMENT[4]

If you are a Brain in a Vat (in short BIV), you are no better than a vegetable but your brain has been technologically managed to make you feel and think that you are absolutely normal. You have no perceptual evidence of looking at a piece of paper but that is one of the important things that matters for your knowing that you are reading a paper. You cannot take help from the public because each of them could very well be a BIV. The BIV argument differs from the Evil Genius argument on the ground that the former is not only worked out on a logical possibility but also on a physical possibility. In view of the technological advancements, it is possible that the brains are in vats and connected with supercomputers which are so advanced that each of us feel as if we are all BIV. The BIV Argument can be put forth as

A6:
P1: If you know that you are reading a paper you know that you are not a BIV.
P2: You do not know that you are not a BIV.
Therefore, you do not know that you are reading a paper.
Why should you know that you are not a BIV if you know that you are reading a paper? The answer (given below as A7) presupposes the validity of Knowledge Closure.

A7:
P1: You know that you are reading a paper.
P2: You know that if you are reading a paper then you are not a BIV.

Therefore, you know that you are not a BIV.

So far we discussed that the principle of Knowledge Closure is presupposed by different arguments for skepticism.

4. H. Putnam (1981), "Brains in a Vat", in his *Reason Truth and History*, Cambridge Cambridge University Press.

Now we switch on to the Zebra that refutes these arguments for skepticism by refuting the Principle of Knowledge Closure.

II

Dretske's Zebra and Two Criticisms

DRETSKE'S ZEBRA[5]

Fred sees a zebra in a zoo. Fred's eyesight is alright, he is at a close distance from the animal he is looking at. Fred has sufficient evidence that justifies him to hold the belief that the animal in front of him is a zebra. In fact, it is a zebra. Thus Fred knows that it is a zebra. Secondly, that z (=It is a zebra) entails that m (= it is not a mule cleverly presented as a zebra) is a known entailment for Fred. However, Fred is not justified in believing that m, hence, Fred does not know that m. It becomes a counter example to Knowledge Closure. Fred knows that z and knows that z entails m but does not know that m. In other words, knowledge is not closed under known entailments. It is not necessary that knowledge of p gives rise to knowledge of q, even if p entails q is known entailment of a subject S. Although the Truth-operator holds good with respect to

A8:
P
P entails q
Therefore, q

in the sense that if the premises are true, the conclusion is true too, Knowledge-operator does not hold good. We find counter examples to

5. F. Dretske (2005), "The Case against Closure", in M. Steup and E. Sosa (eds.), *Contemporary Debates in Epistemology*, Malden, MA: Blackwell Publishing Limited, pp. 13-26. .

A9:
Known that p
Known that p entails q
Therefore, known that q

And, thereby, we find counter examples to the principle of Knowledge Closure. This is Dretske's standpoint expressed through the zebra example.

Why is it that Fred does not know that m? The set of evidence sufficient to justify Fred's belief that z is not sufficient to justify Fred's belief that m. The m-evidence is not contained in the z-evidence. For example, Fred has not checked or verified the honesty and sincerity of the zoo authority for the belief that z and yet justified in believing that z; but in not verifying that Fred is not justified in believing that m.

If the zebra case is alright, KC is refuted and, thereby, skeptical arguments presupposing KC are refuted.

Your belief that a (= you are actually seeing me) is justified even if your belief that d (= you are not dreaming now) is not justified and you know that a entails d and know that a.

You are justified in believing that t (= 2+2=4) even if you are not justified in believing that e (= you are not deceived by the Evil Genius) and you know that t entails e and know that t.

You are justified in believing that p (= you are reading a paper) even if you are not justified in believing that v (= you are not a BIV) and you know that p entails v and know that p.

KLEIN'S CRITICISM[6]

Although the z-evidence does not contain the m-evidence, the z-evidence plus Fred's knowledge that z entails m does

6. P.D. Klein (1981), *Certainty: A Refutation of Skepticism*, The Harvester Press Ltd., UK.

justify Fred's m-belief. In other words, refutation of KC does not follow from the fact that Fred has the justified z-belief and an unjustified m-belief, but only if it so happens when Fred's knowledge of the entailment is included under the evidence for Fred's m-belief. Inclusion of this known entailment as evidence, according to Klein, does make Fred justified in believing that m.

STINE'S CRITICISM[7]

Fred's m-belief needs no evidence once Fred knows that z. Dretske is right in claiming that the m-belief is not a relevant alternative that blocks the knowledge of z. But, the exclusion of m-belief from the set of relevant alternatives does not prove that Fred does not know that m. On the contrary, it proves that the knowledge of m is not to be claimed, in normal circumstances and, hence, not needing evidence. The z-belief presupposes the m-belief in a manner that makes it misleading to claim the knowledge of m when the knowledge of z is claimed. It would be like claiming that my wife is either in the kitchen or in the bedroom after claiming that my wife is in the kitchen, although, literally, the former is true if the latter is true. In normal circumstances, it is odd to claim both.[8]

III

The Tautology Case

I propose the following counter example to KC and show how it overcomes the kind of criticisms made by Klein and Stine.

7. G. Stine (1976) "Skepticism, Relevant Alternatives, and Deductive Closure", *Philosophical Studies,* 29 (1976), pp. 249-61. Reprinted in K. DeRose & T.A. Warfield (eds.), *Skepticism: A Contemporary Reader,* New York: Oxford University Press (1999), pp. 145-55.

8. This example is in Klein (1981), p. 152; originally given in H.P. Grice, (1961), "The Causal Theory of Perception", *Proceedings of the Aristotelian Society,* Supp. 35.

Imagine that in a question-answer session, you are asking and I am answering questions in the following way:

Do you know that you do not know every tautology?
Yes.
Do you know that 2+2=4?
Yes.
Do you know that every tautology is implied by any proposition whatsoever?
Yes.
Do you know that even if you do not know a tautology t, any proposition does imply t?
Yes.
Do you know that even the proposition 2+2=4 implies that t?
Yes.

Now, if I am right in answering "Yes" in each case, and I think I am right, then,

I know that 2+2=4, know that t is implied by 2+2=4 and yet I do not know that t.

Thus, the principle of Knowledge Closure fails.

How do I know that 2+2=4 implies t, without knowing t? Of course, I do know many implications without knowing their consequents. For example, I know that if you are a bachelor then you are an unmarried man but I do not know that you are an unmarried man. The question is not on knowing of any implication but on knowing of an implication in which the consequent is a tautology. It boils down to the question, how do I know that t is a tautology without knowing that t? Because, if I do not know that t is a tautology then I do not know that t is implied by 2+2=4, and it seems to be absurd that I know that t is a tautology without knowing that t is true.

My answer to the above question is very simple. I need not know every tautology in order to know that I do not know every tautology. Secondly, I need not know every tautology to know that every tautology is implied by 2+2=4. Therefore, I need not know t to know that I do not know t. And, I need not know t to know that 2=2=4 implies t. Even if I don't know t, I know that t is implied by 2+2=4, as much as I know that you are a bachelor implies that you are an unmarried man even if I do not know who you are and whether you are an unmarried man or not.

To get the situation clearer, let us imagine that we are playing a game of noting down tautological sentences. The rule of the game is that he/she will be the winner who writes more number of tautological sentences than the other within five minutes. I do not know what you wrote as to be your 2nd tautological sentence but do know that 2+2=4 does imply that sentence. You might not have written a tautological sentence at all. But that is out of context. If we are playing the game and it is a fair play of the game, then, no doubt you have written a tautological sentence and I do know that you have written a tautological sentence even if I do not know what exactly the sentence that you have written is. It is not a question in the game but outside the game if you question me how I know that your 2nd sentence is a tautological question and hence 2+2=4 implies your 2nd sentence, although I do not know what exactly your 2nd sentence is. It is a question on the fair play of the game, not in the game that has been played in a fair way. The fact remains that I do not know your 2nd sentence but do know that 2+2=4 implies your 2nd sentence and, of course, know that 2+2=4. Consequently, the known antecedent of a known implication is there with the unknown consequent; hence, the principle of Knowledge Closure fails.

The Tautology Case is not defeated by Klein's criticism, nor is it defeated by Stine's criticism. Klein's criticism applied to the case would run like this. Even if what justifies my belief that 2+2=4 does not contain what justifies my belief that t is true, what justifies my belief that 2+2=4 plus my knowledge that 2+2=4 implies t does justify my belief that t. But, in the Tautology Case, the question is not on whether I am justified in believing that t is true. It is rather on whether I do know that t. Clearly, I do not know that t. It is one of my unknown tautologies. Nevertheless, I know that 2+2=4 implies t and that 2+2=4. Therefore, even if knowledge of the implication is included as an evidence along with the evidences for the knowledge of the antecedent, that is not sufficient for my knowledge that t.

Stine's criticism employed to the Tautology Case would mean that evidences for the knowledge of t is not required since the claim of the knowledge of t is irrelevant when it is known that 2+2=4. But this is certainly not the case. The claim of knowing 2+2=4 does not make the claim of knowing t odd.

IV

Conclusion

To conclude this essay, I try to answer two questions. One, why should we make a case against skepticism by countering the KC principle and not by some other means?[9] Two, how is the Tautology Case different from the Zebra Case?

The first question is partially answered in my exposition of the Dream Argument, Evil Genius Argument and BIV Argument. As these arguments can be taken up as some of the significant arguments for skepticism, and these arguments presuppose the validity of KC principle, refutation of

9. Thanks to Madhuchhanda Sen for raising this question.

skepticism can be significantly worked out by a refutation of the KC principle.[10] There is another pressing reason. Such a move has certain advantages over those famous ones like the Cartesian Cogito, Moore's "These are my two hands" and Putnam's proof that "I know that I am not a BIV." The Cogito is a case for certainty based on the logic of assertion, if not the logic of intuition. "These are my two hands" is also a case for certainty based on the logic of what common sense must accept. The advantage of countering the KC principle is that it does not presuppose the certainty of knowledge. Of course, Putnam does not make a case for the certainty of knowledge. He makes a case for externalism. A countering of KC does not require a proof for externalism or internalism.

In the Tautology Case, the evidences for the knowledge of the antecedent (i.e. 2+2=4) and evidences for the knowledge of the consequent (i.e. the unknown tautology, t) are not compared to establish that the latter knowledge is absent in face of the fact that the former knowledge is present. That is, even if the evidences for knowing that 2+2=4 do not contain evidences for knowing that t, this is not the reason why t is unknown to me. On the other hand, in the Zebra Case, m is not known because the z-evidences do not contain the m-evidences. Secondly, the Zebra case is directly a case against the closure of Justification (based on evidences) and, then,

10. Dretske goes to the extent of saying that the KC "pattern of argument is sprinkled throughout the epistemological literature. Almost all skeptical objections trade on it." (F. Dretske, "Epistemic Operators"', in DeRose & Warfield (1999), p. 135.). DeRose (1999) says, "Closure Principle has come to seem as underwriting skepticism" (p. 13). Recently, Dretske (2005) has made explicit his standpoint in favour of denying skepticism by denying KC. He argues that there is no philosophically interesting option but either "we retain closure and accept skepticism" or "we preserve ordinary knowledge by denying closure."

against the KC with the supposition that failure of KC follows from failure of closure with respect to any necessary condition of knowledge. But the Tautology Case is directly against the KC.[11]

11. Warfield (2004) argues against the idea that failure of KC follows from the failure of closure with respect to any necessary condition of knowledge. See T.A. Warfield (2004), "When Epistemic Closure Does and Does not Fail: A Lesson from the History of Epistemology", *Analysis*, 64, pp. 35-41. Brueckner's response (A. Brueckner (2004) "Strategies for refuting closure", *Analysis*, 64 (2004), pp. 333-35.) to Warfield is not against the main contention of Warfield but does make a point of devising a strategy to relate failures of KC with failures of closure for any necessary condition of knowledge. Murphy (2006) (see P. Murphy, "A strategy for Assessing Closure", *Erkenntnis*, 65 (2006), pp. 365-83) provides a better strategy in this regard.

3

Naiyāyika Externalism and Gettier-like Counter Examples

C.A. Tomy

INDIAN epistemology makes a distinction between two types of cognition: *pramā* and *apramā*. *Pramā* is an awareness episode produced by one of the *pramāṇas*, the reliable mechanisms that generate awareness episodes that are true. Cognitions belonging to *pramā* category are bound to be true. An *apramā* cognition, on the other hand, is an awareness episode whose truth is not guaranteed. For instance, *apramās* such as doubt (*saṁdigdha*), hypothetical argument (*tarka*), dreams, memory (*smṛti*), and so on may or may not be true. What are the conditions under which an awareness episode is attributed *pramatva* or knowledgehood? This has been a hotly debated issue in Indian epistemology. My attempt in the present paper is to explicate the conditions of *pramā* (knowledge) for Nyāya school of philosophy. The emphasis would be on those conditions that justify the truth of the awareness episodes considered as *pramās*. Philosophers debate as to what constitutes a suitable justification for *pramā*. Philosophers, known as internalists, argue that an appropriate justification is one made from the point of view of the knowing subject whereas for the externalists justification lies in factors external to the subject. I shall argue that, for a Naiyāyika, to be considered as an instance of *pramā*, a cogntive episode must satisfy the conditions of justification that are both internal and external. Though Naiyāyika epistemology is externalist in its orientation, it acknowledges the internalist moorings of the concept of *pramā*. To establish this point it is necessary to analyse and understand the Naiyāyika concept of *pramā* to which we shall turn next.

I

The Definition of Pramā

Pramā has been defined by the Naiyāyikas as *asaṁdigdha yathārthānubhava* by which they mean that *pramā* is a definite and assured true experience or awareness of objects. On this definition, an awareness episode is considered as *pramā* only if it satisfies the following three conditions:

1. It must be a presentation (*anubhava*) of things,
2. It must true or unerring (*yathārtha*), and
3. It must be indubitable and assured (*asaṁdigdha*).

For a proper understanding of this definition it is necessary to grasp what the Naiyāyikas mean by the key terms *asaṁdigdha*, *yathārtha*, and *anubhava.*

Anubhava is that awareness episode which presents things that are cognized. For an awareness event to be considered as an *anubhava*, its objects or things that are experienced must exist at the time when it occurs. Accordingly, a perception, for example, is an *anubhava* because the object that is perceived actually exists at the time of perception. An *anubhava*, in this sense, is considered as a presentational awareness episode. In other words, *anubhava* is always an experience of things or facts that are given and not those that are imagined or constructed by the mind. In an *anubhava*, the object of awareness episode and the awareness episode in question are causally related. Whenever the object of cognition is present, it immediately and without any time gap produces in the subject an awareness episode which is called *anubhava*. That there is no time gap between the presence of the object and awareness episode is often expressed by saying that an *anubhava* episode is presentational.

A memory episode, on the other hand, is not an *anubhava* because its object may not be present at the time of its occurrence. Even if the memory episode is caused by its object, it is not necessary that the object which causes it should exist at the time of occurrence of the memory episode. In memory, object is not cognized directly and immediately as in the case of perception. For this reason, a memory episode is not considered as presentational but rather as representational in nature.

Anubhava is an internal cognizing event that takes place within the knower and, therefore, it is an internal factor necessary for any knowledge episode.

A Naiyāyika, however, does not think that all *anubhava*s or presentations of things are instances of *pramā*. To be considered as *pramā* an *anubhava* episode must possess an important property namely, *yathārtha*. The word *yathārtha* means things as they are and accordingly a *pramā* episode is an experience of things as they are. An illusion may be thought of as an *anubhava* as there is presentation of things. But in an illusory experience, the objects of experience are not presented as they are. So they are not instances of *pramā* despite there being an *anubhava*. In illusion, they are not presented as they are but as something else. But in a veridical perception things are experienced as they are. Veridical perception is a case of *yathārtha anubhava* because it presents the nature of the things. It should however be noted that all perceptual experiences cannot be considered as *yathārtha*. Naiyāyikas identify two stages of perception, namely *nirvikalpaka* and *savikalpaka*. In both stages of perception, there is presentation of objects. But in *nirvikalpaka* perception we only know that something is present; we do not know the nature of the thing present. Hence, *nirvikalpaka* perception cannot be included in the class of *pramā* episodes even though there is no error involved here as in the case of illusory experience. The nature of the object, that is, the object as it is, is known only in the *savikalpaka* perception, and therefore, when we say that veridical perception is an instance of *yathārtha anubhava* what we refer to is *savikalpaka* perception.

In an *yathārtha anubhava* since the objects of experience are presented as they are, *yathārtha anubhava* could be understood as *true* awareness episode, that is, an experience that corresponds to things as they are. In *yathārtha anubhava,* there is an implicit claim regarding the nature of things that are present. Therefore, we could say that a *yathārtha anubhava* is a true awareness episode that is judgemental in nature. If an awareness event is to be considered as *pramā,* there must be objects of awareness or a state of affairs corresponding to the awareness. The state of affairs or the objects of awareness that make the awareness episode true lie external to the subject. So the *yathārtha* condition is an objective condition that makes an episode of cognition a case of *pramā.*

An *anubhava,* even if it is *yathārtha,* need not be a case of *pramā.* Even when a subject perceives an object rightly if she doubts the existence of the object being perceived or if she entertains doubts regarding the nature of the objects presented, then the internal awareness episode is not attributed *pramatva.* It is one of the conditions of *pramā* that the subject is certain or sure about the truth of the awareness episode. So for a Naiyāyika, the third condition to be satisfied by an awareness event to be qualified as *pramā* is that the truth of the awareness in question is indubitable and assured (*asaṁdigdha*). In other words, a *pramā* episode is one that should not be accompanied by another awareness episode that doubts its truth. Indubitability or non-dubiousness, being psychological in its nature, is an internal condition.

Prima facie, the above account of *pramā* looks radically different from the tripartite definition of knowledge in the Western tradition. The standard formulation of the tripartite definition of knowledge is that S knows that p iff

1. p is true,
2. S believes that p, and
3. S is justified in believing that p.

The first difference that strikes one is that in the Naiyāyika account truth is attributed to a particular awareness episode or *anubhava*. In contrast, in the tripartite analysis, truth is attributed to an abstract entity, namely, a proposition. This difference may not be as radical as it would appear if *yathārtha anubhava* is construed as true judgemental episode as we have done above. After all, concept of a proposition is a later addition in the Western thought. So the attribution to judgemental episode instead of a proposition is not different in spirit. Second, the knowledge state according to the tripartite analysis is a sub-class of belief states. A belief state could be understood either as a dispositional state or an occurrent state. As a dispositional state, a belief state tends to produce either verbal or non-verbal behaviour. Dispositional states are unconscious states. If a belief state is taken as an occurrent state, it is possible to construe belief as a mental act or as an awareness episode. For a Naiyāyika, *anubhava* is not a behavioural disposition but a mental act or awareness episode. This would mean that if a belief is construed as an occurrent state or a mental act, a Naiyāyika would not have any problem accepting a belief episode as an *anubhava* though of course for him occurrent belief states would not exhaust all of one's *anubhavas*. Thus inclusion of belief as one of the components of knowledge in the tripartite analysis does not give enough reason to think that the Naiyāyika analysis of *pramā* moves in a direction contrary to the tripartite analysis.

What is strikingly different in the analyses provided by the two traditions is the third condition that each of them provides. The third condition of the Naiyāyika states that to be considered as *pramā* an awareness episode must be indubitable in the sense that a subject should not entertain any doubt regarding the truth of the awareness state in question. On the other hand, the third condition of the tripartite definition requires that the subject must be justified in holding the belief. But it can be shown that the indubitablity condition of the Naiyāyika analysis can be taken care of by the second condition of the tripartite definition provided belief is taken as a mental act of believing a proposition as true. On this interpretation, the belief condition may be understood as "S believes that p is true."

If S believes that p is true, then S does not doubt the truth of p. Thus the belief condition is to be understood as containing the *Naiyāyika asaṁdigdha* condition of *pramā*. In fact, there is no need to state the *asaṁdigdha* condition separately. For an example, in a formulation of the tripartite definition of knowledge that Gettier attributes to Ayer, the belief condition is replaced by what may be called the certainty condition, namely, "S is sure that p is true."

This formulation appears to be closer to the Naiyāyika point of view. (This suggests that there is perhaps no need to state either the belief condition or the indubitability condition. Both can be taken care of by the certainty condition.)

If the indubitability condition can be reduced to the belief condition as we have seen above, then the tripartite analysis seems to have a condition that Nyāya does not possess namely the justification condition. But is it true of Nyāya that it does not subscribe to the justification condition? If this is so, then any awareness for a Naiyāyika, any true awareness episode would be a *pramā* thereby making the Naiyāyika conception of knowledge weaker than the tripartite notion of knowledge. However, that the Naiyāyikas are silent about the justification condition does mean that they do not subscribe to it. It does not require much effort to realize that justification is already built into the Naiyāyika concept of knowledge, for, like any other schools of Indian philosophy, Nyāya conceives *pramā* as true awareness derived from or presented by *pramāṇas*, the reliable sources of knowledge. Justification for *pramā* consists in the fact that a *pramā* awareness is caused by a *pramāṇa*. Though various schools of Indian epistemology disagree among themselves as to the number of acceptable *pramāṇas*, each of them believes that in order to consider a true awareness state as *pramā* it must be presented by one of those mechanisms which the school accepts as reliable.

The goal of the third condition is to appropriately connect the internal factor (awareness episode) with the external factor (the truth or the state of affairs). The justification states why awareness must be taken to be true, and in this sense the function of justification is to guarantee the truth of awareness in question.

In the light of the above discussion, we can say that from the Naiyāyika point of view, *pramā* can be defined as an *indubitable, justified, true awareness episode* where awareness episode is understood as judgemental in nature. This understanding of *pramā* is not radically different from the tripartite analysis of knowledge as justified true belief provided belief is understood not as a behavioural disposition but as a mental act in which the subject is certain about the truth of the proposition believed and the awareness episode is not restricted to acts of belief.

The Naiyāyika definition of *pramā* should be acceptable to each school of Indian epistemology. However, they disagree with regard to the way truth of an awareness episode is revealed. The Indian internalists, *svataḥ prāmāṇyavādins*, argue that if an awareness episode possesses truth, its truth would be self-revealing. For this purpose, it does not require the aid of any external factor or mechanism. The Indian externalists, called *parataḥ prāmāṇyavādins*, on the other hand, argue that an awareness episode does not have the power to reveal its own truth by itself. The revelation of the truth of an awareness state requires the satisfaction of certain external conditions. This points to the grave differences between the internalists and externalists with regard to the nature of the justification for the truth of the awareness. The internalist argues that the justification has to come from within the knowing subject. For the externalist, on the other hand, justification comes ultimately from sources external to the subject. This has implications for the conceptions of truth that these two camps advocate. For an externalist, truth is basically a relation of correspondence between the awareness episode and the state of affairs that is its object. But for an internalist, truth is a property of an awareness state internal to the subject or it emerges from the way an awareness state is related to various other cognitive episodes of the relevant types.

II

The MīMāṁsaka Internalism

Though my primary aim in this paper is to explicate the Naiyāyika version of externalism, before I move on to do this

I shall very briefly discuss internalism as it would help us understand much better the Naiyāyika externalism when it is compared and contrasted with the internalist position.

For an internalist, or a *svataḥ prāmāṇyavādin*, to be considered as *pramā*, it is sufficient that the awareness is produced by a *pramāṇa*. Hence, the only justification for the truth of *pramā* episode consists in the fact that the awareness in question arises from *pramāṇa*. There is no need for any additional factors to turn an awareness episode into a knowledge episode. As a result, for the internalist, justification is purely internal. This view is based on the assumption that an awareness event arising in a subject tends to be a knowledge episode. This tendency is present in an awareness episode even when the awareness episode in question is infected with doubt, because the function of doubt is to take us to certainty or "the non-dubious awareness of truth."[1] Since the justification for the truth of the awareness episode is internal, as soon as the true awareness episode arises, its truth is revealed to the knower without any support from external factors. For this reason, knowledgehood or *pramatva* of an awareness episode is revealed by itself (*svataḥ*). For the internalist, as Matilal puts it, knowledgehood or *pramatva* is "an intrinsic, natural, non-derivative or unconditional property"[2] of awareness episode. Knowledgehood is revealed to the awareness event as soon as it arises by itself.

If there is a natural and intrinsic tendency of every awareness episode to be a *pramā*, then it would follow that an awareness event cannot be false. However, a number of our

1. B.K. Matilal, "Knowledge, Truth and Pramatva", in B.K. Matilal, *Mind, Language and World* ed. J. Ganeri, Oxford: Oxford University Press, 2002. pp. 149-61.

2. Ibid., p. 154.

awareness episodes including those derived from *pramāṇas* as evidenced in the case of perceptual errors are false. Therefore, an internalist must explain how an awareness episode that arises from *pramāṇas* can go wrong. Mīmāṁsakas who are *svataḥ prāmāṇyavādins* argue that if an awareness event fails to be knowledge, it is because of extrinsic conditions that inhibit its attainment of *pramatva* or knowledgehood. As Maṇḍanamiśra puts it: *Prāmāṇyam svataḥ; aprāmaṇyam parataḥ.* That is, in the absence of extraneous factors working negatively, the knowledgehood of an awareness is revealed in and by itself. On this theory, a claim to knowledge can be refuted only by pointing out the external factors responsible for the error or defects in the causal mechanism that produced the awareness.

Any account of knowledge is incomplete without specifying how one comes to know that one has knowledge. The knowledge that one knows — second order knowledge as it is often referred — is very important from the point of view of the knowing subject. Because without knowing that one knows, one cannot make any knowledge claim from a first person point of view. Discussion about the nature of second order knowledge in Indian epistemology is centred around three seminal questions:

1. Once a cognitive episode has occurred, how does one know that it has occurred? Or, how is the occurrence of a cognitive event revealed to the subject?
2. If the cognitive event in question is a knowledge event, how does one come to know about its knowledgehood or *pramatva*? In other words, how does one know that the cognitive event, assuming that it is a knowledge event, is indeed a knowledge event?
3. Does the process or the event that reveals the

occurrence of the cognitive event also reveal its knowledgehood if it is a knowledge event?

Mīmāṁsakas agree among themselves with regard to their answer to the third question: the very same process or the event that reveals occurrence of an awareness event reveals its knowledgehood in case the awareness event in question is a *pramā* episode. However, the answers given to the first and second questions by various schools of Mīmāṁsā differ radically. The Prābhākara school argues that once awareness state arises in the subject, it reveals itself as an awareness state. Since knowledge is an intrinsic property of the awareness event, when the awareness episode is revealed, its knowledgehood is also self-revealed. This shows that, for Prābhākara, there is no distinction between first order knowledge and second order knowledge. No knowledge event can occur without its knowledgehood being revealed. A knowledge event is self-revealing. Therefore, to say that one has unrevealed or unconscious knowledge is a self-contradiction.

The Bhāṭṭas as against the Prābhākaras hold that when a cognitive event occurs, it cannot reveal itself to the knower. An awareness event by itself is imperceptible to the subject. Its occurrence is inferred from the "cognizedness" of the object that the subject is aware of. If the awareness event in question is a knowledge event, the very same process of inference that reveals its cognizedness reveals its knowledgehood as well. Thus the occurrence of an awareness event as well as its knowledgehood is revealed to the subject by a process of inference.

Murārī Miśra, also a Mīmāṁsaka, is of the view that neither the Prābhākaras nor the Bhāṭṭas are right about second order knowledge and the way an awareness episode is revealed. He agrees with the Bhāṭṭas that an awareness episode cannot reveal itself to the subject and that another episode is required for its revelation. He, however, disagrees with the Bhāṭṭas that it is through a process of inference that a cognitive episode is revealed to the subject. Rather an awareness event is exposed to the subject by an internal perceptual event that follows it immediately. This internal perceptual event, called *anuvyavasāya,* also reveals the knowledgehood of the awareness if the awareness episode belongs to *pramā* category.

III

The Naiyāyika Externalism

The Naiyāyikas who are *parataḥ prāmāṇyavādins* argue that if an awareness episode is true, its truth is not self-revealed but is made manifest by certain external conditions. As a result, an awareness episode by its nature is not endowed with *pramatva;* rather it acquires its *pramatva* because of suitable (*guṇa*) extraneous factors. Of course, to be considered as *pramā,* an awareness state must be derived from the *pramāṇas* for they are undoubtedly the reliable mechanisms that provide us with knowledge. But the production of awareness by *pramāṇas* by themselves does not guarantee its truth as evidenced in the case of, for example, perceptual errors. Therefore, there is a need for some additional factors that would convert an awareness event into a knowledge episode. In other words, an awareness event is a *pramā* episode only if it gains external support. So knowledgehood is a contingent property of awareness and not a property that an awareness episode is entitled to by its very nature as the Mīmāṁsakas would hold.

The first and foremost external factor that turns an awareness episode into knowledge is the presence of a fact corresponding to it.[3] This is what makes an awareness episode

3. B.K. Matilal, op. cit., p. 155.

true. But its truth again is not sufficient to make it a *pramā* episode. The awareness event and the fact must be appropriately related. So the Naiyāyikas must specify the nature of the relation between the true awareness episode and the fact corresponding to it. For the Naiyāyikas an awareness episode, like any other episode, is an effect (*kārya*). A *pramā* episode being an awareness episode is an effect. This means that *pramā* is a causal product and this view opens up an avenue for a causal analysis of *pramā*.

Matilal advocates a causal analysis of *pramā* which, according to him, means that a true awareness is causally produced by the *pramāṇas*. He is of the view that in the case of *pramā*, evidential justification coincides with justification in terms of causality.

In fact, the notion of *pramāṇa* as the "causal instrument" to *pramā* or knowledge event has this dual character. It is both causal and evidential. Justification of a perceptual knowledge (this is a chair) is accomplished by citation of its causal connection with the sense object contact. Although it is not the same cause as the presence of chair in the room causes my perceptual knowledge of it, it is the evidence of the senses which justifies its knowledge claims as well as accounts for its origin.[4]

An evidential justification is one made by the subject on the basis of factors available to the subject's perspective or internally and directly accessible to him. This means, evidential justification is purely internalist in nature. For the belief that p, the evidential justification a subject might provide would be something like "I have seen that P," "I know that such and such is the case and from that I inferred that P", etc. Matilal's

4. B.K. Matilal, op. cit., p. 152.

point is that the evidential justification like these can be made in causal terms because *pramāṇas* are the causal instruments to *pramā*. That is, instead of expressions like "I have seen that p" or "I have inferred that p" one can justify one's awareness episodes by saying "My visual perception has produced my awareness that p" or "That my awareness that p is the causal product of my valid inference." Thus *pramāṇas* justify the truth of an awareness and accounts for its origin.

Matilal's aim in providing causal-evidential account was to extend the *pramāṇa* theory by including causal account and thereby establish the Naiyāyika externalism unequivocally and on a firm foundation. However, his interpretation of *pramāṇas* as "instruments" that cause and provide with evidential justification appears to be neutral in the sense it is consistent with both the Naiyāyika externalism and the Mīmāṁsaka internalism. So if the causal-evidential account of knowledge, as understood by Matilal, is accepted, then it would follow that there is no difference between the internalist and the externalist accounts of justification for *pramā*. Moreover, an account of *pramā* on this line would give rise to skepticism. The Gettier-like counter examples that Śrīharṣa raises against the Naiyāyika account of knowledge must have been motivated by an account similar to the causal-evidential account of justification as formulated by Matilal.

Śrīharṣa formulated a number of counter examples to dispute the Nyāya-Vaiśeṣika definition of knowledge as non-dubious true experiencc:

1. While gambling a person makes a guess which happens to be true. The gambler was absolutely certain about the outcome. But can we say that the gambler's awareness was a case of *pramā*?
2. A person sees a cloud of dust on a distant hill, but he

mistakes it for smoke. On the basis of this misperception or the false evidence provided by the senses, he concludes that there is fire on the hill. As a matter of fact it happens that there is fire on the hill which makes his awareness episode true.

3. From a distance, a person sees two horns and infers the presence of a bull in the field. But later it is discovered that he sees a pair of wooden horns. The real horns of the animal were cut-off and in its place artificial ones were fitted a long time back. In this case, the subject's awareness about the presence of bull was true and there was evidential justification for the truth of the awareness. Śrīharṣa's point is that though all the conditions of knowledge are satisfied, this would not be considered as a case of knowledge.

Each of these cases satisfies the definition of *pramā* as given by the Naiyāyikas and yet, argues Śrīharṣa, these are not instances of knowledge. The point that Śrīharṣa wants to drive home is that despite having justification, the awareness episode in each of these cases under consideration turns out to be true only accidentally. As justification in the above cases does not ensure the truth of the awareness episodes, truth is not properly and adequately connected to the awareness episodes. Truth being only accidental to the awareness episodes under discussion, there is always the possibility of error.

Matilal tries to overcome the difficulties posed by Śrīharṣa by drawing on the distinction between the first order knowledge and the second order knowledge implicit in the Naiyāyika account. Such a distinction is available also to the Mīmāṁsakas except the Prābhākaras for whom there is no distinction between the first order and the second order knowledge. Unlike the Prābhākaras, the Bhāṭṭas and Murārī Miśrā believe that the occurrence of a cognitive episode can be revealed only by an awareness episode other than itself, and in case the former is a *pramā* episode its *pramatva* too would be revealed by the latter. We have already seen that Murārī Miśrā and Bhāṭṭas disagree with regard to the nature of internal episode that reveals a cognitive event. For Muārī Miśrā it is internal perceptual event called *anuvyavasāya* whereas the Bhāṭṭas believe it to be an inferential episode. The Naiyāyikas along with the Bhāṭṭas and Murārī Miśrā reject the Prabhākara view that a cognitive episode by itself can reveal its own occurrence as well as its knowledgehood. Further, the Naiyāyikas do not agree with the Mīmāṁsaka position that the very same internal episode that reveals the occurrence of a cognitive episode can reveal its knowledge-hood. First of all, the occurrence of a cognitive episode cannot be revealed by itself. On this matter, they are in perfect agreement with Bhāṭṭas and Murārī Miśrā. The second, the same episode that reveals the occurrence of a cognitive episode having the *pramatva* cannot reveal its *pramatva*. In order to reveal *pramatva* of a cognitive episode, there must be an internal awareness episode other than the one that reveals its occurrence. Thus when a cognitive episode occurs, to reveal its occurrence, there must be a second episode, an episode other than itself. Similarly, if the original cognitive episode is a *pramā* episode, to reveal its *pramatva* a third episode, an episode other than the one that reveals its occurrence is required. Moreover, the nature of episode that reveals the occurrence of the original cognitive episode and that of the one that reveals its *pramatva* are different. The Naiyāyikas believe the occurrence of a cognitive episode is revealed by an *anuvyavasāya*, an inner perceptual episode that immediately follows it. If the initial awareness episode is a knowledge episode, then the Naiyāyika position on the knowledge of knowledge is that neither the initial awareness nor the awareness that immediately follows it reveals its knowledgehood. The knowledgehood of the initial awareness is revealed to the knower by a process of inference from a reliable evidence. The occurrence of a knowledge episode (first order knowledge) does not necessarily imply the occurrence of a second order knowledge episode, just as the possession of a pot of gold does not imply that the possessor knows that he possesses a pot of gold. The reason for this is that the knowledge episode and the awareness episode that reveals its occurrence and the awareness that manifests its knowledgehood are three distinct awareness episodes. These three episodes may not even be simultaneous or concurrent. So, a Naiyāyika can grant unrevealed or unconscious knowledge. A knowledge episode may occur without being revealed. Similarly, a knowledge episode can occur without its knowledgehood being revealed or known.

Matilal calls the first order knowledge episode "knowledge simpliciter" and the awareness episode that reveals its knowledgehood: "Knowledge of the knowledgehood of the knowledge simpliciter."[5] According to Matilal, this distinction can help us overcome the skeptical attack on the Naiyāyika account of knowledge by separating the subjective and objective aspects involved in the occurrence of a *pramā* episode. One of the ways in which a skeptic attacks knowledge is as follows: when a subject makes a knowledge claim, he does not have any other reason to appeal to, except his subjective certainty. But even when the subject is absolutely certain about the truth of the awareness, he could be mistaken. So certainty cannot be an adequate ground for knowledge.

5. B.K. Matilal, op. cit., p. 153.

On the basis of the distinction between knowledge simpliciter and knowledge of knowledge, the Naiyāyikas can maintain that a knowledge event can occur without revealing its knowledgehood. An occurrence of a knowledge event may be revealed to a subject without revealing its knowledgehood for the process by which the occurrence of an awareness is revealed is different from the process by which its knowledgehood is revealed if it is a knowledge episode. For example, when one entertains a doubt about the truth of an awareness event, one is aware of the awareness episode without being aware of its knowledgehood. Similarly, one may be sure about the knowledgehood of his awareness on the basis of some evidence. But subsequently when further false evidences are furnished, one one may give up his certainty of knowledgehood of the awareness. This happens even when the awareness in question continues to be true. This further shows that awareness which is knowledge simpliciter and the awareness of the knowledgehood of knowledge simpliciter are two distinct awareness episodes produced by different causal processes. This distinction, argues Matilal, is useful in resolving skeptical puzzles of the sort that Śrīharṣa raises.

Despite having knowledge, a subject may refuse to make a knowledge claim when evidences counter to the knowledge claim are presented to them. In such cases, though the subject possesses knowledge simpliciter, he does not have knowledge of knowledge simplicter. Thus, in all the three cases that Śrīharṣa describes, the subject knew but the subject did not know that he knew. This strategy of resolving Śrīharṣa's puzzle consists in adopting a third person approach to knowledge and jettisoning the subjective or psychological aspects in the conception of knowledge.

The Navya-Nyāya straegy that Matilal adopts and advocates is undoubtedly in tune with the Naiyāyika externalism. It could, however, be seen basically as a desperate attempt to minimize the damage already done by Śrīharṣa. In fact, it concedes that Śrīharṣa is partly right; he is right about the first person account of knowledge. Matilal's charge against Śrīharṣa is that he failed to appreciate the third person perspective on knowledge. Though the strategy is appealing from an externalist point of view, it is counter-intuitive at least for three reasons.

First, since true awareness is defined as knowledge, it would imply that there is no distinction between truth and knowledge. Any awareness episode that is true would pass for knowledge. Matilal is aware of this but believes that though it makes the position a bit uneasy, the move makes great strides in tackling Śrīharṣa's counter examples. Therefore we stand to gain in the trade off. Second, the standard adopted for judging knowledge and knowledge of knowledge are different. In order to consider an awareness event as knowledge simpliciter, it is sufficient that it be true. But in the case of knowledge of knowledge, in addition to the truth of the second order awareness, there is an additional requirement that there be sound evidence. This is because the second order knowledge is the result of an inference. Matilal argues:

> . . . the second order awareness or knowledge of knowledge should be treated differently. Knowledge of knowledgehood is an inference. False or inadequate evidence for such inference affects the conclusion. That is, we cannot accept this conclusion as knowledge, even if this is a true conclusion, when the evidence adduced or used is shown to be false or inadequate.[6]

---

6. B.K. Matilal, op. cit., p. 161.

In other words, as far as second order knowledge is concerned, the awareness episode in question should be true, and justified by a sound inference whereas in the case of knowledge simpliciter, it is sufficent that the awareness state is true. Third, certainty is important for proper knowledge because it is quite counter intuitive to say "I know but I do not know that I know." In other words, for an adequate account of knowledge proper, first person account is as important as the third person point of view. We may say that we have knowledge only when the first person and third person accounts match or are consistent. A subject must be able to encash his knowledge but this would not be possible without certainty from the part of the subject. A knowledge event must lead to successful action. Thus lack of psychological certainty would go against one of the conditions of knowledge put forward by the Naiyāyikas themselves.

Is there a way to avoid these problems and come up with an account of knowledge that is not affected by Śrīharṣa's criticism? In my opinion, such an account is available to the Naiyāyikas. Matilal was almost there when he attempted a causal analysis of knowledge. But his account failed because, for him, causal account justification and evidential justification amounted to the same. If the causal account can be properly supplemented, it is possible to come up with an account that is not liable to the skeptical attack of the sort that Śrīharṣa raises. The key to this lies in the version of the causal account of justification that Goldman has provided in order to overcome the Gettier counter example to the tripartite definition of knowledge.

According to Goldman, for the justification of a subject's belief that p, there must be a causal link between the fact that makes p true and the subject's belief of p. Accordingly, for him, "*S knows that P* if and only if *the fact P is causally connected in an 'appropriate' way with S's believing that P*."[7]

In other words, in order that a belief state may be considered as knowledge, it should be caused by the very same state of affairs that the belief state represents. In this case, the initial cause of the belief does not lie in the senses or the evidence provided by the senses but in the external world. What is important to be noted about Goldman's causal account is that he stipulates that the causal connection between the fact that make a belief true and the belief itself must be an appropriate one. By an appropriate causal connection between them, he does not mean that the fact that p directly causes the belief that p, there could be intermediary causal links. For Goldman, the appropriate sorts of causal connections include perception and memory. There could be causal connections other than perception and memory and in such cases, the appropriateness of the causal connection lies in that the subject is able to properly reconstruct the causal chain from the fact that p to subject's belief that p. The subject's reconstruction is proper only if the relevant causal links actually occurred the way the subject believes them to be.

Goldman's causal account of knowledge goes well with the Naiyāyika externalism, except that Naiyāyikas would not accept memory as providing appropriate causal links. They would readily agree that the causal links extend beyond the evidence provided by *pramāṇa*s and would argue that the evidence provided by the *pramāṇa*s must be appropriately and causally connected with facts or objects in the world. In other words, there has to be a proper causal relationship between the evidence from which we have inferred the fact and the

7. A.I. Goldman, "A Causal Theory of Knowing", *Journal of Philosophy* 64 (1967), pp. 357-72; see p. 359.

fact itself. Such a causal chain rules out the possibility of an awareness event becoming accidentally true.

Once Goldman's version of the causal account of justification is incorporated into Naiyāyika externalism, the Gettier-like counter examples that Śrīharṣa raises would not hold any longer against the Naiyāyika concept of *pramā*. Śrīharṣa's gambler does not have knowledge despite the fact that his awareness episode is true and that he is absolutely certain about its truth because the awareness episode is not caused by the fact it is about. In fact, a Naiyayika does not even have to appeal to the causal factor in this case; he would not consider guess as a valid source of knowledge. In the second case, the presence of fire is inferred on the false evidence that there is smoke on the hill. We need to note two things here. First, the sensory evidence that there is smoke on the hill is not caused by smoke. Second, the "smoke" in question was not produced by the fire present on the hill. Therefore, the subject's perception of *vyāpti*, here, namely, the invariable relation between smoke and fire, is based on a misperception and is therefore not justified in inferring fire on the basis of the invariable relation between smoke and fire. The premises involved in the inference are products of misperception and are, therefore, false. As a result, the inference is not a sound one. Therefore, the awareness of fire though true would not be an instance of *pramā*. Moreover, the invariable relation between smoke and fire must be understood in causal terms. That is, the cloud of smoke in question must have been produced by the fire on the hill. If the smoke on the hill is produced in the kitchen in the valley and blown up to the hill, then the inference to the effect that there is fire on the hill is not sound and is not knowledge. In short, as the true awareness state and the fact are not suitably connected by cause-effect relation, the subject does not possess the knowledge that there is fire on the hill. In the case of inference, we should be able to reconstruct the causal links. If the subject were to reconstruct the causal links, the reconstruction would not be a proper one as some of the subject's beliefs about the causal links would not be true.

The case of the bull with artificial horns is more sophisticated; we might be tempted to say that the subject knows that there is a bull in the field because there is a causal link between the presence of the bull and the appearance of artificial horn. We might say that artificial horns were fitted on to the animal some time in the past and because of this act there is a causal link between the presence of the animal and the preference of artificial horns. There is no denying that there is a causal link. But this causal link is something that takes place quite accidentally. There is no lawful causal relation between the presence of the bull and the presence of the artificial horns. In the language of Nyāya, there is no *vyāpti* between these episodes. As a result, the subject is not justified in inferring the presence of the bull and hence the subject does not know. (Though the inference in question may be said to be valid, it is not a case of knowledge as it is not a sound inference. In the case of knowledge based on inference, it should be a necessary condition that the inference must be sound. Soundness of the inference would rule out the possibility of its conclusion being true accidentally.) The subject's perception of *vyāpti* in this case is on mistaken grounds, the subject's reconstruction of the causal links between the facts and the awareness episode involve false beliefs.

Though the causal external condition of justification plays a very important ingredient in the conception of *pramā*, it by itself does not constitute a sufficient justification for knowledge. For a proper account of knowledge, there must be an additional condition of justification which is internal.

This internal condition, as we have noted already, states that the subject should not have any doubt regarding the truth of the awareness episode. The internal condition of justification — let us call it the "certainty condition" — helps us overcome those counter examples that cannot be taken care of with the causal external conditions described above.

Consider the Gettier counter example that Harman raises. Harry sees Tom steal a book from the library and testifies before the university judicial council that he saw Tom steal a book from the library. Now Tom's mother testifies that Tom has an identical twin called Buck. She admits that Buck has a bad character but hopes that he did not steal the book. She, however, lies to the council that Tom was thousands of miles away when the theft took place. Neither Harry nor the members of the council know that she was lying. The question is whether Harry can, under these circumstances, claim to know that Tom stole the book. Of course, Harry was initially certain about the truth of his belief that Tom stole the book and his belief was indeed true. However, because of the false evidence provided to him, he is no more certain. Now Harman argues that under these circumstances Harry can no longer claim to know that Tom stole the book.[8]

Matilal argues that it is counter-intuitive to hold that Harry did not any longer know when false evidences are provided to him, because his awareness that Tom stole a book from the library continues to be true and caused perceptually by the actual event of Tom's stealing the book. However, to say that Harry knew was also counter-intuitive in some sense. Matilal's way out of the dilemma consists in distinguishing between

8. G. Harman, "Selection from Thought", in *Essays on Knowledge and Justification*, eds. George S. Pappas, Marshall Swain, Ithaca: Cornell University Press, 1978, p. 212.

knowledge smpliciter and knowledge of knowledge. That is, Harry knows that Tom stole the book but does not know that he knows. So it would seem that Matilal's strategy to deal with various formulations of the Gettier counter examples is to switch back and forth between knowledge of knowledge and knowledge simpliciter.

In my opinion, Matilal's way of dealing with the problem does not do justice to the Naiyāyika account of *pramā*. Harman's intuition that if a subject is not certain about the truth of the proposition that he believes, then the subject does not possess knowledge is one which a Naiyāyika would readily accept. For them, if the certainty condition is not met, then the subject does not possess knowledge. So with regard to Harman's counter example, the Naiyāyika would maintain that the subject knew that A stole the book. But when doubt arose because of the false evidence, he ceased to possess knowledge.

At this point it is quite legitimate to ask how a knowledge episode can lose its *pramatva* or knowledgehood in the presence of false evidence. This question arises from the mistaken assumption that knowledgehood is a property of a single mental state or an awareness episode. Knowledge-hood is not to be construed as a property of a single atomic awareness episode rather it is the property of a group of awareness episodes which are related with one another in a certain way. If the knowledge episode in question is empirical in the sense that it pertains to the external world, then within the complex knowledge episode, there is at least one atomic episode which is causally related to the world or the state of affairs it represents. The relational treatment of knowledge within the Naiyāyika framework would imply that a complex awareness episode that has *pramatva* or knowledgehood would consist of the awareness of evidence, awareness of the fact for which we have evidence, the awareness of the awareness of the evidence from which one infers the presence of the fact, and finally the awareness that one has knowledge that is inferred from the awareness of the evidence and the awareness of the fact. And one of the constituents of this complex awareness episode, namely the awareness of the fact, is caused in an appropriate way by the fact that makes it true. This account shows that causal justification of knowledge and coherentism can go hand in hand and not mutually exclusive as is often thought of. *Pramatva* of complex awareness episode emerges when both internal and extenal conditions of knowledge are satisfied.

On the interpretation that is pursued in this paper, the Naiyāyikas put forth a very stringent set of conditions for attributing knowledgehood. In order to consider a (complex) awareness episode, both internal and external conditions are to be satisfied. The internal awareness episode must be true and it must be caused by the state of affairs it is about and that the subject must be certain about the truth of the awareness in question. This conception of knowledge is very close to our commonsensical notion of knowledge and satisfies our philosophical intuitions concerning knowledge. Above all, it successfully overcomes Gettier counter examples which appear in various colours and flavours.

4

Matilal's Analysis of Contemporary Indian Approaches to Epistemology

K. Srinivas

I

EPISTEMOLOGY is one of the oldest branches of philosophy that is basically concerned with the scope and aim of knowledge, the means of knowledge, and the justification of knowledge. There are instances in the Western tradition where philosophy is reduced to epistemology. For example, the Continental rationalists and the classical British empiricists found nothing more enterprising than epistemology in their respective doctrines. Similar is the case with metaphysics. For example, in the philosophies of Hegel and Heidegger the focus on metaphysics is conspicuous. But there are also systems or schools of thought that have a balanced view as regards the status of epistemology and metaphysics for they believe that these two traditional branches of philosophy are not independent from each other, rather they are interdependent. Therefore, they cannot afford to ignore one or the other. The classic examples of this approach, there may be many, are Kant in the Western tradition, and Indian philosophical systems in general. Of course, in the philosophy of Nyāya the epistemological and logical concerns dominate any other aspect of philosophy. This does not mean that Nyāya did not subscribe to any metaphysical standpoint. As an allied system of Vaiśeṣika, Nyāya admitted its metaphysics. Now we have, at least, two distinctive approaches to epistemology. They are: (1) epistemology *per se* is independent of any other branch of knowledge; and (2) epistemology cannot be separated from metaphysics. The Western epistemologists in general subscribe to the former view, and their Indian counterparts to the latter. For my present study I have chosen one of the most distinguished contemporary Indian philosophers, late Professor B.K. Matilal. I propose to discuss his approach to epistemology in general.

In his book, *Perception: An Essay in Classical Indian Theories of Knowledge*, Matilal writes that:

> The dispute that lasted in a little over twelve centuries between the Nyāya and the Buddhist over the nature of perception, the critique and criteria of knowledge, and the status of the external world is undoubtedly an important chapter in the history of global philosophy.[1]

Just as we come across realism, phenomenalism, and representationalism in the Western epistemology, we also come across these doctrines in the Indian philosophical systems, especially in Nyāya and Buddhism. Therefore, for any Western reader, as rightly pointed by Matilal, this would be a familiar area. Another interesting point is that some modern Indian philosophers have produced illuminating expositions and reformulations of some of the speculative metaphysical doctrines with a view to highlight the traditional style of philosophizing in India. This is a laudable effort on the part of those who have undertaken this exercise of making

1. B.K. Matilal, *Perception: An Essay on Classical Indian Theories of Knowledge*, Oxford: Clarendon Press, 1986, p. 1.

the philosophical theories of India more prominent and appealing. Yet, the modern Indian philosophers, who are under the influence of analytical tradition of the West, may not find these doctrines all that illuminating. The reason that is attributed to this state of affairs by Matilal is that these writings are largely opaque and blurred. I do not quite agree with Matilal's stand on this issue. Of course, Matilal himself held that most of these metaphysical theories were discussed out of context. Consequently, these most sophisticated theories were conveniently ignored as inessential.[2] I personally feel that the Western epistemologists have to reconcile with the fact that these theories are not understood by the modern Western philosophers not because they are opaque and blurred, but because their method of analysis cannot penetrate into the structural depths of the classical Indian philosophical theories. I would say that the method of analysis advocated by the modern Western epistemologists is found wanting in many respects. To put it in the language of Wittgenstein, the rules of the game played by the classical Indian philosophers are totally different from those of the game played by the modern Western philosophers.

It is opined by Matilal that *pramāṇa* method, to use Indian terminology, is invariably linked up with the *prameya*s. To quote Matilal in this context:

> The metaphysical doctrines of classical India developed in the background of intense intellectual activity in philosophy. Hence they are imbedded in the philosophical style that was current at that time, and this was subtly oriented by epistemological concerns and orchestrated by the logical theories of classical India.[3]

2. B.K. Matilal, op. cit., p. 8.
3. Ibid.

Matilal was right when he held that the establishment of any metaphysical theory is dependent on certain epistemological considerations (*pramāṇādhīna prameya-sthitiḥ*). To put it in the modern terminology, our epistemological and logical concerns precede our metaphysical and ontological ones, for the latter can be approached only with the help of the former.

II

In order to drive his point home, Matilal highlighted the importance of *pramāṇa* epistemology and its logical significance in determining our metaphysical/ontological[4] commitments or priorities. He assumes that all of us, when we are in a pre-philosophical state, believe that there is a world external to us, and it exists independent of us. This is a position taken by realists in general. However, it is not all that easy to philosophically prove or disprove this commonplace belief. For centuries together philosophers have advocated theories after theories either to establish or to falsify this commonplace belief. The validity of pre-philosophical, pre-reflective, and pre-critical certainties is always doubted by philosophers. The commonplace belief that there is a material or physical world independent of our awareness of it is questioned by philosophers by advancing and vindicating a counter-thesis that the so-called external reality or world is only mind-dependent. The skeptics on the other hand viewed it in a different way. According to them, it is not possible to arrive

4. Normally it is held that metaphysics deals with the nature of highest reality, which may be transcendental. Those who are averse to using this expression, especially those who claim to be anti-metaphysicians in the West like Marxists, the modern empiricists (logical positivists), preferred the term ontology to metaphysics to deal with the ultimate nature of reality, which is not transcendental. But, there are philosophers who interchangeably use these expressions to deal with the ultimate constituents of reality.

at any definite position as regards the status of material or physical world. All that philosophers could do is to advance the theories of various sorts, which often conflicted with each other, to explain their position as regards the status of the material or physical world. Therefore, these theories are highly presumptuous. They are only poor substitutes for truth. This being the case, suggests Matilàl, "if the clothes do not fit we may either decide to remain naked or buy new ones that may fit better. The skeptics may prefer the pristine purity of nakedness, but others choose the latter."[5]

It is generally held that when our beliefs are proven to be true they acquire the status of knowledge. One of the paramount interests of philosophy as an activity is to define the characteristics of knowledge, and to set forth certain criteria for obtaining knowledge in the sense in which its characteristics are defined. In this sense, claims Matilal, the purpose of epistemology or *pramāṇa-śāstra* in Indian tradition is not in any way different from that of the Western tradition. Epistemology, which is basically concerned with theory of knowledge, investigates and evaluates the evidence, our methods of reasoning, and the criteria upon which our knowledge claims are dependent.[6] Apart from that, the common point shared by both Indian and Western philosophical traditions is, as aptly put forth by Matilal, their concern for truth or reality. Therefore, the quest for knowledge in both the traditions has given rise to a number of interesting results. According to Matilal, *pramāṇa* is that which leads to a knowledge episode called *pramā* as its end.

Interestingly, evaluation of our evidence to knowledge is invariably linked up with the question of the sources of

5. B.K. Matilal, op. cit., p. 21.

6. Ibid., p. 22.

knowledge. Out of these sources, one very important source recognized by the empiricist tradition in the West is sense experience, which provides us with immediate knowledge. Accordingly, it is held that the real base for our theoretical and objective knowledge lies in observation of aspects of reality. In its weak form it is held that all our knowledge must start with sense experience and its validation is always subject to some form of observational data. Some *pramāṇa* theorists in India have advocated the weaker version of the empiricist tradition of the West. One of the advocates of such a position is Uddyotakara. According to him: "we emphasize perception, for all *pramāṇa*s (in some way or other) [are] preceded by (sensory) perception."[7] The result of such a position is that conception without perception is empty. The counter-thesis of this claim is that perception without conception is blind. But Matilal claims that this counter-thesis too is incompatible with empiricist doctrine, for it advocates the view that pre-conceptual or non-conceptual perception is blind and unrevealing. However, such a non-conceptual experience is possible as advocated by the Indian epistemologists in the form of *nirvikalpaka pratyakṣa*. But such a possibility does not deter the *pramāṇa* theorists from advocating their theory. In fact, such a perception maintains the ontological neutrality regarding the status of concepts.

Further Matilal maintains that the most important aspect of such a philosophic position is "experience." Both our factual knowledge and knowledge of existence are necessarily justified on the basis of experience. If this were so, we need to define the expression "experience" in clear terms. Both the Western and Indian epistemologists are not very clear about this expression. Matilal feels that the most suitable Sanskrit

7. B.K. Matilal, op. cit., p. 22.

expression for the expression "experience" is *pratīti* or *anubhava*. Like in the Indian tradition, in the Western tradition too the expression "experience" is used as an ultimate court of appeal for any knowledge claim. However, this argument does not stand the critical scrutiny for obvious reasons. The most immediate, non-conceptual experience is barren because it does not have any explanatory content within it. Consequently, nothing can be put forward in the form of a legitimate statement.[8]

III

Matilal analyses the various theories of perception of classical Indian type in the context of *pramāṇa* epistemology. For this purpose first he highlights the salient features of *pramāṇa* doctrine that inevitably resulted in a form of skepticism. One of the chief traits of skepticism is that it always questioned the very possibility of our knowledge of the external world. Thus it remained a constant shadow of epistemology both in Indian and Western traditions. In a way, it has driven the subject of experience to think how the world "appears" to the subject. Thus it paved the way for a distinction between an experience and its interpretation, between a crude sensory data and logical construction out of them.

Matilal argues that empiricists claim sensory experience provides us with the "building blocks" of knowledge. If what is called knowledge is constructed out of these "building blocks," then the very edifice of our knowledge represents mere appearance, but not the reality. Such a treatment of knowledge made a skeptic to contest the legitimacy of the empiricist claim to the knowledge of the external world, for it is not established. Even the causal theory of perceptual experience is no better for it makes the subject of experience

---

8. B.K. Matilal, op. cit., p. 23.

speculate about the true nature of the external world which is never given in perceptual experience. According to Matilal, both the Western epistemologists and the *pramāṇa* theorists are sailing in the same boat when they try to look for, with a sense of hesitation, some subjective element in support of our objective knowledge. Now it is left for the epistemologists of both the traditions to bridge the gap between subjective and objective elements to substantiate our knowledge claims. *Pramāṇa* theorists in general hold the view that no knowledge is possible independently of some perceptual basis or other. Even the scriptural knowledge is not exempted from this condition. They hold the view that direct experience of such people like the Buddha or the Jaina resulted in scriptural knowledge. It is believed by the Naiyāyikas that the Vedas are spoken by God. Hence, their validity is as sure as the validity of the statements made by a reliable and trustworthy person. Although Mīmāṁsā does not support this claim, the Vedānta joins the camp of *pramāṇa* theorists in this regard.

When Matilal speaks of empiricism, he uses this expression with utmost caution. He does not use it as a movement that opposed the continental rationalism of the West. Following the flag of W.V. Quine, Matilal holds that an empiricist doctrine is one that insists on the observational basis of our knowledge. However, there may be philosophers who do believe in the existence of mind-independent universals or realities. Such a claim to mind-independent realities does not really bar them from being empiricists. The sixth-century philosopher Bhartṛhari held that all of us possess certain amount of innate disposition to articulate concepts in speech. This is an acquired disposition, maintains Matilal, for it is derived from (in a metaphysical sense) the residual (memory) traces of the experiences of every individual in countless previous births (*pūrvahita saṁskāra*). This view of Bhartṛhari can be compared with that of Plato. According to the latter, learning in a way is a process of recollection of that knowledge possessed by an individual before birth. The Buddhists claim that universals are mere convenient myths. Like a vehicle they are dispensable once the goal is achieved. On the contrary, Nyāya holds that some natural universals are objectively real and perceivable for they have instantiations that are perceptible. The point that Matilal wants to drive home here is that although there are doctrinal differences among the various schools of Indian tradition, yet majority of the schools can be brought under the camp of *pramāṇa* theorists.

IV

Skeptical arguments against accepted doctrines are found both in the Western and the Indian traditions of epistemology. The *pramāṇa* theorists in general have their own arguments against the skeptics' claim that knowledge is a remote possibility. According to Nyāya epistemology, perhaps one of the earliest arguments against the challenges of skepticism is that we can know what is out there with the help of *pramāṇa*. Therefore, the object of knowledge (*prameya*) is definitely known. Before the emergence of full-fledged philosophical systems in India, there were ancient skeptics like Sañjaya and others (between 600 to 100 BCE) who directed their skepticism against the possibility of knowledge pertaining to moral, religious, and eschatological matters. Sañjaya questioned the veracity of the statements made about these matters. According to Matilal, the following questions never found any satisfactory answers. Therefore, they remained unexplained. They are as follows:[9]

1. Does anything survive death?
2. Is the world finite?

---

9. B.K. Matilal, op. cit., p. 27.

3. Is there a soul different from body?
4. Does he who acts "enjoy," i.e. get reward and punishment?
5. What is right and what is wrong?

The skepticism about the truth claims of the answers to the above-mentioned questions gradually paved the way for the skepticism about the truth claims of the very possibility of knowledge. It is in fact in the writings of Nāgārjuna (second century CE), the great Mādhyamika Buddhist, one can come across a systematic and skeptical challenge to theory of knowledge. Later on it was the skeptics like Jayarāśī (eighteenth century CE) and Śrīharṣa (eleventh century CE) who questioned the claims to knowledge. There were also the skeptics in the form of monistic metaphysicians who were critical of *pramāṇa* method. Thus there was a constant debate between the advocates of *pramāṇa* theory and the skeptics.

The skeptical dialecticians of India followed a "radical" method to expose the knowledge claims of the epistemologists. They contested that the very concept of knowledge and its foundations are either paradoxical or circular. Matilal defends the Indian skeptics like Nāgārjuna, Sañjaya, Jayarāśī, and Śrīharṣa for they have not indulged in the construction of any metaphysical system. Although it is true of Sañjaya and Jayarāśī, it is not the case with Nāgārjuna and Śrīharṣa. Matilal admits that it is arguable whether Nāgārjuna and Jayarāśī have subscribed to any metaphysical position at all. Perhaps, the skepticism advocated by them is only complementary to their soteriological goal in the sense that it serves as a ladder to climb up, and then to be discarded. The aim of these two skeptics is to show the limitations of pure reason in sorting out the knowledge claims of one sort or other. In the process of resolving the rival knowledge claims pure reason would lose itself in the quicksand of contradictions. In fact it is held in the *Laṅkāvatāra-Sūtra*:

> The own nature of things cannot be ascertained by the analytical exercise of intellect (*buddhi*). Therefore, they (the things) are shown to be ineffable and 'without own nature.'[10]

This statement suggests that reason cannot discover the real nature of things. In a way, it is a reminder to the rationalists that reason cannot penetrate into the very nature of things.

Matilal believes that philosophical empiricism is ingrained in some form of skepticism. The very problem of uncertainty regarding the claims to knowledge necessitated epistemologists of both traditions to look for secure foundations of knowledge that rest on direct evidence (experience). The *pramāṇa* theorists subscribed to only a weaker form of empiricism. According to them, sense experience is only a starting point of knowledge. The reason is that most of them advocated other sources of knowledge apart from sense experience. On the contrary, the skeptics held that there is an inherent contradiction between the data offered by experience and reason. Such a contradiction always leads to skepticism but not to any legitimate theory of knowledge. For instance, Nāgārjuna put forth the following argument to justify his claim that there is no possibility of obtaining knowledge. The paraphernalia that we use to measure or obtain knowledge is itself in need of justification. If this point is conceded then the question of arriving at knowledge of any kind is a remote possibility for this process ends up in an infinite regress. Thus Nāgārjuna's argument remains a great impediment to *pramāṇa* theorists for knowledge in any form. If this argument is not conceded then our choice of selecting a particular standard of measurement

---

10. *Laṅkāvatāra-Sūtra*, tr. D.T. Suzuki, London: George Routledge and Sons, 1932, ch. II, verse. 173; ch. X, verse 167.

is completely arbitrary and not really appealing to reason. A Cartesian might say that one can choose a standard of measurement that is highly indubitable and self-evidently true. But the skeptics like Nāgārjuna would object to this proposal, for the search for an indubitable criterion necessarily involves subjective element into the whole process. This may not be acceptable to those epistemologists who insist on some objective standards of criteria.

In order to defend their position, the *pramāṇa* theorists, in spite of their different ontological commitments, would agree that whatever exists can be known. Further, they argue that whatever is known is existent. Consequently, it is effable in the sense that it is expressible or nameable in language. Of course, there are some *pramāṇa* theorists who hold the view that knowability is not a sufficient condition for effability. This is not treated as a major impediment. According to Bhartṛhari, human consciousness always reveals the objects (knowables) with the help of words for it is ever vibrating. In the absence of such a mechanism, the revelatory character of human awareness is destroyed. Perception or awareness of any sort would be meaningless in the absence of word-mediated act of consciousness. Nyāya does not subscribe to this view for it admits perception or awareness without effability. This is due to the fact that there are objects or knowables that are only known or revealed to human consciousness, but may not be expressible through the medium of language. To use Wittgenstein's terminology, they can be called "mystical".

According to Matilal, the thesis of "ineffability" may be viewed from different standpoints. According to one interpretation, what is grasped by the senses (perceptual experience) cannot be put into words or language, for language is a social affair and it can only communicate those aspects of reality that are inter-subjectively accessible. Since perceptual experiences are unique to each subject, the content of such experiences cannot be communicated. Matilal holds that such a view can be, with minor modifications, attributed to the Buddhist doctrines advocated by Diṅnāga and Dharmakīrti. Both these philosophers have subscribed to what is called phenomenalism and logical atomism in modern terminology. Just as Russell held that there must be logically proper names in any ideal language, although one cannot give an example for it, Diṅnāga too held that the pure particular or pure sense-datum cannot be, in principle, communicated. Thus it remains ineffable (*anirdeśya*).[11] The other interpretation is holistic in its nature. Accordingly, it regards the reality as a single indivisible whole, which is undifferentiated. Since the whole cannot be grasped in its fullness, language reduces it into parts resulting in the proliferation of concepts. Consequently, all our concepts enter into some contradiction or other. As long as language operates with the help of concepts, it fails to picture reality.

V

Another interesting feature that we come across in the empiricist tradition of the West is the cleavage made between a priori (non-empirical) and a posteriori (empirical) forms of knowledge. The former is the knowledge obtained independent of experience and the latter is necessarily dependent upon sense experience. Such a distinction is not found in the Indian epistemological tradition. This may be treated, according to Matilal, as a lacuna in the *pramāṇa* theory advocated by the Indian epistemologists. For the sake of convenience, Matilal classifies knowledge into non-empirical and empirical forms. The scriptural knowledge or the knowledge of *dharma* can be characterized as non-empirical

11. B.K. Matilal, op. cit., p. 30.

knowledge for it cannot be derived through any empirical means such as perception or inference. Here the expression *dharma* is used in its loose sense to cover religious, social and ethical duties of individuals. And we cannot obtain knowledge of these matters empirically. This is the reason why the skeptics like Sañjaya did not entertain any questions regarding these matters for knowledge about these matters is impossible. What Matilal wants to stress here is that the scriptures impart non-empirical or trans-empirical knowledge. *Pramāṇa* theorists recognized this form of knowledge for the tradition defines scripture as follows: "the means (in fact, "knowledge-how") that can be known by neither perception nor inference is what they come to know through scriptures. And this constitutes the scripturehood of the scriptures."[12]

What is unique and significant about the scriptural knowledge is that it is highly infallible. It is held by the great Vedika philosophers like Bhartṛhari and Śaṅkara that the knowledge obtained by empirical means such as perception and inference may be fallible at times. Matilal calls it a kind of fundamentalism for our knowledge derived from the authority of scriptures remains infallible because there is no empirical means that can ever falsify it. To record the statement of Śaṅkara in this context:

> The truth (knowledgehood) of the Vedic statements is self-established independently of anything else. It is like the sun which reveals itself while revealing colours.[13]

Thus scriptural knowledge is self-validating. Another important argument advanced by *pramāṇa* theorists in general is that scriptures being apocryphal there is no human agency involved in their creation. They are the creations of the divine

12. B.K. Matilal, op. cit., p. 32.
13. Ibid., p. 35.

(*apauruṣeya*). If a statement is false, its falsity can be traced back to the shortcomings of those who propounded it. But in the absence of an author it is treated valid. This amounts to a dogma. However, Matilal puts forward the following considerations to avoid the charge of dogmatism.

The *Mahābhārata* and *Manu-Smṛti* held that the truth of *dharma* is beyond human comprehension. However it can be obtained by the means such as scripture, the verdict of the saints and seers, and one's own conscience or moral intuition. It is on the notion of intuition or *pratibhā* that Bhartṛhari has something to say. He clearly demarcated intuition from perception and inference.[14] Intuition occurs to us like a flash. It is altogether a different type of experience. Intuition occurs in all sentient beings in a spontaneous manner. Such a form of knowledge, according to Bhartṛhari, is far more reliable than any other form of knowledge for it is direct and immediate. But Matilal argues that although it is not clear from Bhartṛhari's interpretation of intuitive knowledge whether it can be treated as non-empirical or empirical, it is a clear case of non-empirical form of knowledge. Further Matilal alleges that Naiyāyikas, the most prominent *pramāṇa* theorists, find it difficult to accommodate it in their classification of empirical forms of knowledge. The scriptural knowledge as viewed by Bhartṛhari, is the highest form of intuitive knowledge.

Contrary to the view advocated by Bhartṛhari, Naiyāyikas argued that scriptural knowledge must have some empirical base. They ridiculed the view that the scriptural texts have no author. Just as the veracity of any statement is dependent on the trustworthiness of its author or speaker, the veracity of the scriptural statements must also be dependent on the infallibility of their author. Therefore, according to Naiyāyikas,

---

14. B.K. Matilal, op. cit., p. 33.

the author of the scriptural statements must be a person with perfect knowledge, and he is none other than God. Matilal maintains that the notion of "trustworthy person" can be saved from its sectarian colour and theistic overtone once we accept the Buddha and the Jina as trustworthy (*āpta*) persons. The truth of *dharma* is known through their intuitive insight that is non-empirical in its character. Our religious and moral prescriptions are based on the knowledge of scriptures and the knowledge of *dharma*.

Coming to the empirical foundations of our knowledge, *pramāṇa* theorists hold that all our cognitive knowledge has empirical base. In this context, Matilal cautions us that unlike in the Western tradition, there is no dichotomy made between facts and values in order to understand the basis of our secular, religious, and moral beliefs. In a nutshell, the following are the features of *pramāṇa-prameya* doctrine advocated by the Indian epistemologists.[15]

1. There are valid means or sources of knowledge on the basis of which we make assertions about what exists, and what is true.
2. There are entities called knowables (*prameya*), and each knowable is cognizable. They constitute the world or reality. Each knowable entity can be grasped or apprehended by our knowledge episodes.
3. There is no distinction between means of knowing and its justification. The means themselves justify what is known.
4. *Pramāṇa* theorists are in agreement with the view that like the sensation of pain, knowledge episode is an event.

15. B.K. Matilal, op. cit., pp. 35-36.

5. All knowledge episodes are cognitive episodes, but not all cognitive episodes are knowledge episodes, for some cognitive episodes may not yield truth.

Every *pramāṇa*, maintains Matilal, is both evidential and causal in its nature. It is evidential for it provides ample justification for the cognitive episode. It is causal for it is responsible for any cognitive episode. Therefore it is treated as an "instrument" (*karaṇa*) and most effective causal factor (*sādhakatama*) of the knowledge-episode in question. A *pramāṇa* is that which measures. What is to be measured is called *prameya*. In this sense, *pramāṇa* leads us to such cognitive episodes by means of which we can understand the nature of reality unerringly.

However, reminds Matilal, the skeptical challenge to *pramāṇa* theorists does not end there. Both an empiricist and a *pramāṇa* theorist have to spell out clearly the status and the nature of entities that we experience, and believe to be there. The pre-philosophical or pre-analytical belief in the existence of external world independent of perceiver's consciousness has to be justified. The skeptic questions how a *pramāṇa* theorist can rationally justify his claim that there is an external world independent of perceiver's mind. What are those indubitable data that serve as the foundation of his knowledge claim? If the data obtained from his most immediate experience serve as indubitable foundation of his knowledge claim, then the very nature of that data is under investigation. It is held by the skeptic that the data obtained from one's immediate experience are always subjective, for they are not accessible to others. How can anyone use these subjective data to construct a system of objective knowledge? As rightly pointed by Matilal, there are two alternatives left for a *pramāṇa* theorist to explain away his position. According to the first alternative, he must admit that the objects are exactly the same as those that we experience in our pre-philosophical and pre-reflective mood. Thus the world is constituted by middle-sized, measurable, material objects. The second alternative is that he must construct objects out of the data obtained from his immediate experience to give a philosophical justification to his knowledge claim. These two alternatives led to an age-old controversy between realists and phenomenalists and immaterialists. For instance, Matilal characterizes the Nyāya realism and the Buddhist phenomenalism in the following manner.[16] Following are the features of Nyāya realism.

1. What we directly perceive in our sense experience is the external reality that exists independent of the perceiving subject.
2. All the physical objects, wholes, bodies, and their properties can be seen and touched. We see and touch wholes and substrata because they possess parts and properties, but not because we are able to see and touch these wholes and parts. But in the case of smell, taste, and hear, we do not smell the flower but the fragrance of the flower, we do not taste sugar, but its sweetness, we do not hear the train, but only its whistle.
3. The whole is a distinct entity created by arranging the parts, yet it is not the sum total of its parts. It is over and above the parts. Similarly, a substratum is distinct from its properties it instantiates. In other words, properties subsist in substratum.
4. Either perceiving or seeing is regarded as knowing in the direct sense. There is no secure or basic foundation

16. B.K. Matilal, op. cit., pp. 5-6.

other than perceiving or seeing which is indubitable or certain.

5. Although knowledge is not always verbalized, it is verbalizable.
6. Perceptual illusions can be analysed without bringing in the notion of sense datum or sense impression as an intermediary between the perceiver and the external world.
7. Knowledge in its ordinary sense is neither self-revealing nor self-validating. In other words, it is not necessary that every cognitive event must be noticed or perceived. There is always a possibility that some cognitive events may occur and pass away unnoticed and unperceived. In order to communicate the occurrence of a cognitive event in any form of language it must be inwardly perceived by the subject in question. This inward experience is called (*anuvyavasāya*). One's experience of pain and pleasure comes under this category.

In contrast to Nyāya realism, the Buddhist phenomenalism can be characterized in the following manner:

1. What we are directly aware of in our perception is only a sensible quality, and it is doubtful whether such a quality exists independent of its instantiation. Also, it is doubted whether an instantiation exists independent of our sensation of it.
2. What we call material objects, physical objects, and wholes are perceived only in a secondary or metaphorical sense for we do not have direct access to anything other than their sensible qualities which are directly given in perception. In other words, we

infer the existence of material objects, physical objects, and wholes from their sensible qualities.

3. The so-called material or physical objects are nothing but constructions out of sensory phenomena. A whole is not independent of its parts taken together. The wholes have only nominal existence (*saṁvṛttisat*). They are objects of either unconscious inference or desire-dominated construction. As regards unconscious inference, which is a form of *vikalpa*, it has mere psychological certainty, but not the required logical certainty. Therefore, such inferential awareness cannot be equated with knowledge proper.
4. In its most immediate sense, sensing is identified with knowing. Hence, sensing is the foundation of knowledge. It is indubitable and incorrigible.
5. Such knowledge is not verbalizable, for it is completely free from conception. There is no scope for verbalization without concepts.
6. The entity perceived in perceptual illusion is not distinct from the cognition itself. What is given in sense illusion is an integral part of the sensation itself.
7. Knowledge is always self-revealing. Every cognitive event would be noticed. In other words, no cognitive event can pass away unnoticed. Like pleasure and pain every cognitive event is self-cognized.

The above-cited world-views share the view that the basic elements of our experience are observable or perceptible individuals. But the nature of these individuals is treated differently in both the world-views. According to the realist view of Nyāya-Vaiśeṣika, the observable individuals are physical and the observable phenomena are always explained in terms of the physical. Contrary to this view, the Buddhist phenomenalists hold that there is nothing beyond phenomenal, therefore, everything is reduced to observable phenomena. In spite of these differences, there is another common point shared by both. They view that an ontological system must be properly grounded in epistemology in the sense that what is epistemologically prior must be the staring point of their respective ontologies. For a realist what is epistemologically prior is physical entities with properties. On the contrary, a phenomenalist considers what is directly apprehended in one's sense experience, namely, the phenomenal qualities, become their epistemological priorities.

To conclude: the contemporary Indian approaches to epistemology in many respects resemble the approaches of the Western epistemologists. In fact, it is one's own ontological priority that paves the way for a suitable or convenient epistemological doctrine. It is through the latter that the former gets substantiated or vindicated. The perennial philosophical problems, whether epistemological or metaphysical/ ontological, are common to both the Indian and the Western philosophical systems. The controversy between the realists and phenomenalists as regards the nature of observables in the West is akin to the controversy between the Nyāya realists and the Buddhist phenomenalists in the classical Indian tradition. Matilal consistently highlighted the point that the controversies and issues that dominated the modern Western epistemology are very much found in classical Indian epistemology too. The Western epistemologists cannot ignore the contributions made by the classical Indian philosophers to the traditional theory of knowledge. In the Indian context, epistemology cannot be segregated from metaphysics/ ontology. It is through epistemology that every school of thought justifies its metaphysical presuppositions. In other words, our epistemological concerns are basically rooted in our commitment to a certain type of metaphysics/ontology. Since metaphysical/ontological commitments vary from school to school, there are bound to be differences in their epistemological priorities too. This naturally results in the proliferation of "isms." One cannot deny this fact.

5

Rationality, Thought, and Language

Nirmalya Narayan Chakraborty

IN a paper published in 1950[1] Alan Turing predicted that by the year 2000 it would be possible to build a computer that would have at least 30 per cent chance of fooling an average interrogator into thinking it is a person. Without bothering too much about the prediction, the philosophical significance of the paper lies in the way Turing designs the test, for this very design throws light on the nature of thought that Turing wants to highlight. Turing's opening question was: Can machines think? Setting aside the engineering aspect of the question, the philosophically significant question is about the nature of the claim we make when we ascribe thinking to some creature. What is it that we discover when we recognize that a creature or an object is a thinking being? Aristotle defined man as a rational animal. In order for this definition to be acceptable, this definition must be free from the defects of being too wide and too narrow. In other words, rationality

1. Quoted in D. Davidson, "Turing's Test", *Problems of Rationality*, Oxford: Clarendon Press, 2004. See also D. Davidson, "Rational Animals", in *Actions and Events*, eds. E. Lepore and B. Mclaughlin, Oxford: Basil Blackwell, 1985; D. Davidson, "Thought and Talk", in *Inquires into Truth and Interpretation*, Oxford: Clarendon Press, 1986; D. Davidson, "What Thought Requires", in *Problems of Rationality*, Oxford: Clarendon Press, 2004.

and animality together cannot be present in animals other than humans. Naturally, this would lead to a great debate. Is rationality present in subhuman animals? Even if sub-humans are endowed with rationality, is the difference between human rationality and rationality that is present in subhuman animals a matter of difference of quality or degrees? Related to this question is the issue of the moral relation between humans and animals. One could argue that if animals are bereft of rationality, then humans have a right to use the animals to suit their purpose. Even understanding the nature of intersubjective relation rests on what interpretation of rationality one subscribes to. So it seems that an account of rationality has far-reaching consequences. For my present purpose I would narrow down my discussion to a philosophical account of rationality, i.e. what is involved in ascribing rationality to a creature. Language plays an important role in deciphering one's thoughts and in communication of those thoughts. Language is also an ineliminable element in our understanding of rationality, or so I want to claim. Thus thought, rationality, and language go together. An account of the one would involve account of the other two members in this triad. This is what I call triangulation theory of rationality, thought, and language. It is for you to assess the consequence of triangulation theory with regard to the prediction of Turing or with regard to the nature of computational mimicking of human rationality.

Let us see how we normally use the word "rationality." When we say about a person that she is a rational person, what we want to claim is that she can believe certain things, she can doubt certain things, she can wish certain things, she can imagine, etc. This belief, doubt, wish, etc. are called propositional attitudes. These attitudes always have propositions as their objects. In other words, when we believe we believe something;

when we imagine we imagine something. This something, in philosophical jargon, can be called a proposition. Since these attitudes centre around or grow around a proposition, these are called propositional attitudes.

So to claim that somebody is rational is to ascribe propositional attitudes to her. What is worth noting here is that ascribing a propositional attitude requires ascribing a host of other propositional attitudes. Let us take an example. I say that Sumita believes that Hyderabad is the capital of Andhra Pradesh. This means that Sumita has some beliefs about Hyderabad; she has also some beliefs about Andhra Pradesh; she has some beliefs about what a capital is. Unless Sumita has some beliefs about Hyderabad, Andhra Pradesh and what a capital is like, she cannot be said to have the belief that Hyderabad is the capital of Andhra Pradesh. So ascription of one belief presupposes ascription of a set of beliefs. To determine the content of a person's belief or to identify the belief of a person, we have to take recourse to a set of beliefs that the person entertains. Thus it seems that the identity of a belief depends on other beliefs in the neighbourhood. In other words, understanding or ascription of a belief is possible only against the background of a belief system. And this is possible only because of the holistic structure of beliefs. Beliefs are interrelated in such a manner that talking about one belief leads to talking about other beliefs. So to ascribe rationality to a person is to ascribe belief to her and to ascribe a belief to her is to ascribe a set of beliefs to her. Many animals exhibit through their behaviour that they can make fine distinctions that humans do as well. They recognize individual people or animals, they can find their way back to places where they have been before, etc. But these abilities do not endow them with propositional attitudes. They see and do a lot of things, but they do not know that something is the case.

Further it should be pointed out that the set of beliefs that is required to ascribe a singular belief must be largely true. If the majority of the beliefs of that set are false, then it would not be possible to ascribe a singular belief. Taking our earlier example, if the majority of the beliefs of Sumita about Hyderabad were wrong, if most of the beliefs Sumita has about Andhra Pradesh are wrong, if most of her beliefs about what a capital is are incorrect, then it would not be possible for us to say that Sumita believes that Hyderabad is the capital of Andhra Pradesh. If most of the beliefs that Sumita has about Hyderabad are wrong, then we would say that Sumita does not know which city is called Hyderabad. As a result we cannot ascribe this belief to Sumita that she believes that Hydreabad is the capital of Andhra Pradesh. So in order to ascribe a singular belief to a person, the majority of beliefs that the person holds have to be true. Moreover, if these beliefs lack coherence, if these beliefs are opposed to each other, even then we cannot say that Sumita believes that Hyderabad is the capital of Andhra Pradesh. If Sumita entertains inconsistent beliefs, then again, as before, we have to say that Sumita does not know which city is called Hyderabad. So ascribing a singular belief not only requires ascribing a set of beliefs, also the majority of the beliefs of this set must be true and there must not be any incoherence among those beliefs. This is where rationality lies. We can claim that Sumita is a rational being only when Sumita can be said to satisfy these requirements. Even if there is some incoherence among the beliefs that Sumita entertains, this lack of rationality is explicable only against the fact that majority of the other beliefs that Sumita holds are true and not incoherent. Irrationality is to be explained against the background of rationality. Herein lies the importance of rationality.

Among the different kinds of propositional attitudes we are talking about belief, since belief has a primacy in the set of propositional attitudes. Propositional attitudes like doubt, imagination, wish, etc. depend on belief. If I did not have some beliefs about Hyderabad, if I did not have some beliefs about Karnataka, then I could not even doubt whether Hyderabad is the capital of Karnataka. If I did not have some beliefs about what summer time is like, if I did not have some beliefs about Darjeeling, then I could not wish that I would spend summer in Darjeeling. Thus it seems that primacy of belief has to be accepted. It is in this sense that one could claim that belief is central to thought. While ascribing thought to a person either we ascribe states like belief, imagination, etc. or we ascribe acts like forgetting, etc. And belief demands a special place among these states or acts.

If we want to give a cogent theory of interpretation of human behaviour, we see that there is a relation between thought and language. Thought cannot be said to exist independent of language. Ascription of thought requires ascription of language. Let us explain this further. We can think of two theories here: (1) Theory of Behaviour, and (2) Theory of Linguistic Understanding. Theory of behaviour gives a consistent description of human behaviour. This theory has important applications, for once we are armed with such a theory, we will be in a position to predict human behaviour and that will be important in the field of sociology and economics. When a person decides to behave in a certain way, she has certain beliefs, desires behind her decision. Theory of behaviour aims at giving a consistent interpretation of human decision in the light of her beliefs, desires, etc. This is what can be called a theory of decision. Theory of linguistic understanding explains one's linguistic knowledge, it describes the knowledge possessing which one can claim to know a language. When a person knows a language, she knows the meanings of the expressions of the language. So if we can come up with a theory that gives the meanings of the expressions of the language, then it could be said that knowing a language means knowing the theory. This theory can be called a theory of interpretation.

Now this theory of decision and theory of interpretation complement each other. Theory of decision describes why a person takes a particular decision, how she arrives at the decision, etc. There are certain beliefs, desires that work behind a person's taking a particular decision. Since belief is central to propositional attitudes, theory of decision requires knowing the beliefs properly in order to give an explanation of a person's decision. Now the way to know a person's beliefs is to know which sentences she holds true. If Sumita holds the sentence, "The sun rises in the east," true, then we can say that Sumita believes that the sun rises in the east. So in order to explain decision, we fall back on "holding a sentence true." In other words, if we have a theory that can tell us under which conditions a speaker holds the sentences true, then we can explain the beliefs of that person and as a result we can also explain the decisions of the person. What is to be noted here is that from a theory of a speaker's holding the sentences true, we can get at a theory of interpretation of that language. So to know a theory of interpretation is to know the meanings of the expressions of the language concerned. And once we know the conditions under which the speaker holds the sentences true, thereby we can also know what meanings the speaker attaches to those expressions. If we know that Sumita holds the sentence. "The sun rises in the east" true if and only if the sun rises in the east, we also know what Sumita believes and what meaning the sentence "The sun rises in the east" is assigned by Sumita. Taking the present example, we can say, after looking at the external world, that Sumita believes that the sun rises in the east and we can also tell what Sumita wants to express by the sentence "The sun rises in the east." So here we have simultaneously a theory of decision and a theory of interpretation and we get this from the idea of under which condition a speaker holds a sentence true. From this we can assert that thought (where belief is central) rests on a theory of interpretation. In other words, theory of thought and theory of interpretation are complementary to each other. Ascription of thought requires ascription of a theory of interpretation. One cannot be said to possess thought unless one is an interpreter. If we say that Sumita has thought, i.e. Sumita believes that . . . , then Sumita has to know under which conditions certain sentences are true. In other words, Sumita has to know a theory of interpretation, Sumita has to be an interpreter. And since belief has a holistic structure, Sumita has to know the truth condition of a set of sentences and through knowing the truth condition one can know a theory of interpretation.

To sum up, so far, we cannot ascribe a belief to someone without ascribing a complex set of related beliefs. One cannot be said to have knowledge of the structure of this complex set of beliefs or the relation among the beliefs, if one does not possess the knowledge of language. The complex structure of thought makes the presence of linguistic knowledge essential. Meaning does not depend on the personal whims of a speaker. Meaning resides in a public space. Since ascription of a thought to a person requires ascription of a group of thoughts and ascription of a group of thoughts requires taking help from the expressions (which the person holds true) uttered by the person, the meanings of those uttered expressions do not solely depend on the person. Rather the meanings depend on the linguistic behaviour of the linguistic community to which the person belongs. And a theory of the meanings of the utterances of the speakers of the linguistic community is what is called 'theory of interpretation'. Here we find a close relation between ascription of thought and linguistic understanding. Presence of thought means the presence of the ability to use sentences under appropriate condition. One would fail to make the subtle distinction between different thoughts in a set of thoughts unless one possesses the knowledge of language. So it seems that a creature having thought must be able to possess many thoughts and for this one must be able to understand others' thoughts and also one must be a language user. In other words, a creature possessing thought must be an interpreter. Holistic element in belief relates belief to language in an inseparable way.

Further it is to be noted that in order to have a belief, one is required to have a concept of belief. To ascribe a belief to a person is also to ascribe the concept of belief to that person. To say that 'Sumita believes that . . .' is to say that 'Sumita believes that Sumita believes that. . . .' Unless we think that the person is aware of her own belief, we cannot say that "The person believes that . . .." For the person not only holds one belief, she entertains many other beliefs, since beliefs come in a set. This means that the person is aware of the relation among the beliefs. We cannot ascribe radically incoherent beliefs to the person. We must acknowledge that the person is aware of her beliefs and their logical connections. This just proves that the person is not only aware of her own belief, she has a concept of her own belief. We can analyse the concept of belief with the help of an example. Think of the propositional attitude wonder. I, for example, put my hands in the pocket of my trouser and I was shocked not to find my wallet there. Actually there are two beliefs here. I had a belief (which I had before putting my hands in the pocket) that I had my wallet in my pocket. Later I had another belief (after I put my hands in the pocket) that I had no wallet in my pocket. Just having these two beliefs would not give rise to my propositional attitude of wonder or shock. When alone I am aware of the opposition among these two beliefs, i.e. when I come to know that these two beliefs cannot be true at the same time, I start having the attitude of wonder. So wonder is possible only after there arises a third belief concerning the opposition between the two beliefs. And wonder about at least certain beliefs is both necessary and sufficient condition of thought. Unless one is aware of the coherent relation among the beliefs one holds, unless one holds a belief about whether there is any inconsistency among those beliefs, ascription of thought or rationality to that person would be impossible. Creatures with propositional attitudes and creatures without propositional attitudes might be conditioned to act in a similar way in similar situations. But those who possess propositional attitudes have the ability to introduce a new concept in their conceptual repertoire. For they can be said to recognize the relation among the beliefs in the set of beliefs that they are being ascribed to. Concepts and sentences that cash on them are partly individuated by causal relations with the world and partly by the relations that obtain among themselves. Thoughts, for having propositional content, can only be recognized by finding out the relation among the sentences the speaker holds true. Perhaps by now the reader can understand that my approach to the question whether a creature or a device can think can best be answered by deciphering the structure that thought and language can be said to have.

Now this concept of belief is related to the idea of objective truth or objective world. The fact that I was shocked or I wondered what happened to my wallet implies that there is an opposition between what is true and my earlier belief. What is true does not depend on my believing it or not believing it.

Truth is not subjective; it is objective. As a result, the way the world is does not depend on what my thought or belief is like. For there is an opposition between the way the world is and my belief about it and in that case I accept that my belief is false. A creature that cannot entertain the possibility of being in error, cannot be said to possess thoughts. Once a creature has a concept, it also knows that the concept applies to things independent of what it believes. Thus the possibility of thought depends on the idea of objective truth.

We understand this subjective-objective dualism on the basis of linguistic communication. When I engage in communication with another person, her understanding of the world, her beliefs about the world cannot radically differ from mine. If the worlds of two of us and our beliefs about those worlds were radically different, then linguistic communication could not even take off the ground. If we do not accept that there is a world that is common to us, that this world does not depend on only my or somebody else's belief, linguistic communication would be impossible. So it is through linguistic communication we become aware of subjective-objective duality. If the world were purely subjective, linguistic communication could not take place. And the ability to engage in linguistic communication implies being an interpreter. Thus we can see that rationality, thought, and language are closely related to each other. Language is a social art and so rationality is a communal property. Rationality is born in society, it is placed in society and it is determined by society.

I started this paper by mentioning Turing. Let me conclude my discussion by referring to Turing again. Without getting into a detailed analysis of Turing's test, allow me to make some general comments. The question, "Can machines think?", is different from the question whether a digital computer can pass a certain test. The former requires a criterion of thought against which one can answer the question. In order to do this we have to uncover what goes on when we ascribe thought to a creature. This is what I have tried to do in the previous pages. Whether a particular machine can pass a test depends on the nature of the test, nature of the machine, so on. Though the test can be so designed that it throws light on whether the machine can be ascribed a thought, nevertheless it hardly makes any contribution to our understanding the structure of thought, what possessing a thought requires and this seems to me to be the philosophically problematic question. Once we have a criterion of thought in our possession, we not only can detect whether a creature possesses a thought, we can also determine what the creature thinks. Theory of ascription of thought gives us a theory of content of thought, for identifying the content of thought depends on understanding the relation the particular thought stands with other thoughts in its neighbourhood. This holistic structure of thought is an important element in our theory of ascription.

For argument's sake even if we assume that the creature/ device makes sounds or generates sentences — what seem to be English sentences in response to sentences framed in English — can we say that the creature/device speaks or knows English? The problem is that we do not have any clue to the semantics of the language that the creature/device uses. There is no way that we can determine the relation between the words/sentences that either appear on the screen or are uttered by the creature. What we need is the evidence that the object/creature uses the words to relate to the things or the events in the world, the predicates that it uses in the sentences are true of the things in the world, that it possesses the knowledge of the truth conditions of the sentences it uses. The easiest way for us to acquire the knowledge of these is to see the object/creature interact with the world. Unfortunately this does not happen. In the case of a device its creator can supply the semantics, but then the device can hardly be said to mean something, nor can thought be ascribed to it. For a creature/device to have a semantics it must have causal interactions with the world, and for an interpreter to grasp that semantics she must situate herself in front of the creature/device and must have a common world to share. In short, thought and meaning require a history of a particular sort. This history or origin plays a vital role in understanding one's thought and in interpreting one's language. Any device or any creature that lacks this history can hardly be said to possess thought or express meaning.

6

Knowledge, Belief, and Human Rationality

Smita Sirker

I

The Impetus from Naturalistic Epistemology

HUMANS are epistemic beings. We do possess the indomitable urge to know in varying degrees. We are in constant search to know the limits of our knowledge and seek the means to transcend that barrier. The limit is both internal and external, out there in the world. Three questions appear fundamental: *what is it to "know"?* (i.e., *what is knowledge?*); *what can we know?*; and *how do we know what we know?* Philosophers' interest in epistemology is not just scholastic in nature but it is a much deeper reflection of our basic instinct to know, to enquire about this urge itself. What is knowledge? What can be qualified as "knowledge"? Putting it simplistically, knowledge is information (preferably correct ones) from the external world forming its representations in the human mind of the external world which gets transformed into beliefs about our world and its people. Ultimately knowledge is the corpus of "beliefs" we have about our world. But beliefs are not without qualification. Not all beliefs will qualify as "knowledge." Traditional epistemology had taken up the task of setting standards for the kinds of belief that will qualify as knowledge.

Descartes' epistemological inquiry in the *Meditations* starts with the question: What propositions are worthy of belief? Cartesian epistemology, termed as *foundationalism,* is said to have the following two schemes: to identify the criteria by which we ought to regulate acceptance and rejection of beliefs; and to determine what we may be said to know according to those criteria. Mainstream epistemology dwelled on the problem of identifying the criteria of justified belief and dealt with the sceptical challenge to the possibility of knowledge. The concept which dominated epistemology was that of *justification*.

Human beings are also rational animals. The fact that we are animals places us squarely within the order of nature. But being rational also liberates us to an extent from the absolute rule of all external forces and we are endowed with some measure of self-determination. The possession of rationality, our ability to decide and act on good and sound reasons makes us free agents (to a certain degree). Free will is the capacity to choose and do as we want, either by following the call of reason or by sidelining it. Our constant interaction with the world within and outside us provides with a constant flow of information both in and out of ourselves. Such information with proper refinement becomes knowledge which gradually builds our corpus of beliefs. But are we free to choose and decide on our entire body of beliefs (preferably constituting only of justified/true/rational beliefs)? Our cognitive faculty of reasoning and our cognitive repertoire of beliefs are intertwined in multiple ways than we can think of. The connection between one's rationality (reasoning) and one's belief system may seem obvious to many. But, reasoning is not just about drawing inferences/conclusions/decisions in given situations — be it correct or incorrect, logical or illogical, rational or irrational. It involves cognitive, psychological processes and the outcome of the processes rest heavily, may be primarily, on the repertoire of beliefs that one has. Further, when we evaluate an individual's decision as rational or otherwise — it may appear *prima facie* that it is only the "decision" of the agent that is being evaluated. In actuality, we also look for preceding reasons, causes, or justifications behind the decision made or the action taken. And this eventually and invariably leads us in most cases to look into the gamut of beliefs that the agent possesses.

Now, what kinds of belief do we entertain? It's varied. Most of us possess true beliefs, false beliefs, justified/rational beliefs and unjustified/irrational beliefs. No matter how we identify and qualify them — the facts are: (i) we do possess a mixed bag of beliefs — true and false; (ii) possess beliefs with or without justification; possess rational and irrational beliefs; and finally (iii) we do not possess the absolute control over choosing all our beliefs nor do we have a mechanism to sieve all the justified beliefs into our system and leave out the rest. It is both conscious and unconscious. In fact, many would not mind entertaining unjustified/irrational beliefs because they work just fine in their lives. Rather than being inconvenient, these beliefs often prove convenient. Thus, many of our beliefs fail to pass the traditional JTB (*justified true belief*) criterion of knowledge, and do not constitute the ideal "knowledge" base of the person. Analytical epistemology adopted a normative approach to evaluate our beliefs and test whether they qualify to be justified enough to become proper knowledge. But how often are we actually aware of what constitutes our real knowledge and what does not? Giving normative criterion is often not sufficient to tackle real-life situations. Humans are imperfect epistemic agents, imperfect rational agents — and they are what they are supposed to be in this complex world. It is too utopian to think of perfect rational, epistemic beings who will successfully lead a normatively guided life. The naturalistic move in epistemology opened up an alternative proposal that viewed humans as cognizers with natural limitations and suggested the use of empirical science to study their epistemic activities. Study of epistemology with humans as epistemic agents cannot be theory laden. Such approaches inadvertently develop an explanation gap between real observations and prescribed theories. The naturalistic move stressed on the application of methods, results, and theories from the empirical sciences. It differed from approaches that emphasize a priori conceptual analysis. Though varieties of naturalistic epistemology differ in terms of how they conceive the relationship between empirical science and epistemology, how much they rely on empirical science in theorizing about knowledge, and which sciences they take to be particularly relevant to epistemological questions, these approaches opened up the possibility of empirical studies in philosophy. W.V.O. Quine and A.I. Goldman deserve special mention in this context.

Quine in his essay, "Epistemology Naturalized"[1] argued for conceiving epistemology as a "chapter of psychology" and for seeing epistemology and empirical science as containing and constraining one another. According to him, the foundationalist programme has failed; the "quest for certainty" is a "lost cause." Epistemology's main problem is to explain the relationship between theories and their observational evidence. One way of doing so is the psychological study of how people produce theoretical "output" from sensory "input." There is a legitimate epistemological investigation of providing a scientific account of how, on the basis of "meagre" sensory stimulation, people have such elaborate theories of the world.

1. W.V.O. Quine, "Epistemology Naturalized" in *Ontological Relativity and Other Essays*, New York: Columbia University Press, 1969.

According to Quine, the input consists of physical events, "the stimulation of sensory receptors," and the output is a "theory" or "picture of the world," that is, a set of "representations" of the cognizer's environment. He says,

> The stimulation of his sensory receptors is all the evidence anybody has had to go on, ultimately, in arriving at his picture of the world. Why not just see how this construction really proceeds? Why not settle for psychology?[2]

He continues,

> Epistemology, or something like it, simply falls into place as a chapter of psychology and hence of natural science. It studies a natural phenomenon, viz. a physical human subject. . . . The relation between the meagre input and the torrential output is a relation that we are prompted to study for somewhat the same reasons that always prompted epistemology: namely, in order to see how evidence is related to theory, and in what ways one's theory of nature transcends any available evidence. . . . But a conspicuous difference between old epistemology and the epistemological enterprise in this new psychological setting is that we can now make free use of empirical psychology.[3]

Jaegwon Kim[4] points out that Quine's epistemology and analytical epistemology could be viewed as sharing a common subject matter, namely, they are both concerned about "beliefs" or "representations." The only difference is that while the former investigates their causal histories and connections, the latter concerns itself with their evidential or justificatory

2. Quine, op. cit., p. 75.

3. Ibid., pp. 82-83.

4. J. Kim, "What is Naturalized Epistemology?", in *Philosophical Perspectives* 2, ed. James E. Tomberlin, Asascadero, CA: Ridgeview Publishing, 1988, pp. 381-406.

properties and relations. As Kim says "Quine is urging us to replace a normative theory of cognition with descriptive science."[5] Analytical epistemology studied rationality, justification, belief, and knowledge. The search was for the adequate epistemic grounds — whether we had enough justificatory support for our beliefs about the world. On the contrary, Quine proposed to study the nomological explanation, the causal connection between our sensory evidence and our beliefs about the world. Where traditionalists looked for epistemic support between these two relata, Quine looked for the causal relation between the same relata. Knowledge, thus approached, is a natural phenomenon, the outcome of a natural process whereby sensory stimulation leads to theories about the world. To understand the connection between the sensory stimulation and the theories, and to understand how far beyond the stimulation our theories go, we need to study the process scientifically. The point is not to "reconstruct" the transition, but to understand it as it actually occurs. Quinian naturalistic epistemology is thus "contained" in psychology as a sub-discipline. However, he also says that there is a sense in which naturalistic epistemology "contains" the rest of science. Our theories and beliefs about the world, which constitute our science, are part of epistemology's subject matter. Because they "contain" one another, epistemology and the rest of science can be mutually constraining. This conception of the relationship between science and epistemology contrasts vividly with the traditional view of epistemology as "queen of the sciences."

Unlike Quine, Goldman[6] does not see epistemology as part of science. Instead, Goldman thinks that answering traditional

5. Kim, op. cit., p. 389.

6. A.I. Goldman, *Epistemology and Cognition*, Cambridge, MA: Harvard University Press, 1986.

epistemological questions requires both a priori philosophy and the application of scientific results. As he often puts it, epistemology "needs help" from science. Goldman's theory of knowledge, a form of *causal* reliabilism, holds that a justified true belief counts as knowledge only if it is caused in a suitably reliable way. To be "suitably reliable," a belief-forming process must have a propensity to produce more true beliefs than false ones, and the process's own causal ancestry must have a greater propensity to produce reliable processes than unreliable ones. His theory itself gives empirical science an important place in our understanding of knowledge. The most evident place where psychology matters is in the identification and evaluation of belief-forming processes. It is psychology that tells us what processes cause our beliefs, and it is psychology that enables us to judge their reliability. Thereby the determination whether a particular belief is a case of knowledge will turn on both philosophical and psychological considerations. Philosophically, we can say that the belief (if justified and true) is knowledge if it was caused in a suitably reliable way. Whether actually it was caused in such a way, however, is a question for empirical science.

Goldman's approach to epistemic justification is also reliabilist and grounded in science. It is an important task of epistemology to clarify and describe our *epistemic folkways* — the set of our commonsense epistemological concepts and principles, including the concept of justified belief. We also require a theory of what justified believing is. To study our epistemic folkways, Goldman thinks it is necessary to empirically examine the ways in which we apply and acquire our epistemic concepts. He hypothesizes that we apply epistemic concepts to individual cases in much the same way we apply most of our concepts, that is, by judging how similar or different a particular case is to stereotypical instances of the concept. For epistemic justification we compare the process whereby a person has come to believe something with what we take to be typical justification-conferring processes, such as perception or deduction. He contends that we take such processes to confer justification because we take them to be reliable. Our commonsense understanding of what processes people use to arrive at their beliefs and our commonsense assessments of their reliability are apt to be quite different from the psychological truth of the matter. So, in Goldman's view, it is also necessary to construct a theory of what epistemic justification really is, as opposed to how common sense takes it to be. That theory will be grounded in our psychological understanding of how beliefs are formed, and it will include assessments of those processes in terms of reliability.

Thus, we find that the naturalistic approach to epistemology gave an impetus to take recourse to empirical investigation wherever necessary. Quine and Goldman have strongly argued in support of naturalization though the former insisted on epistemology being a sub-discipline to psychology while the latter advocated "help" from science. Goldman reinstated justification grounded in science. The emphasis on empirical investigation should not be seen as just an alternative suggestion to the conventional methodology of doing epistemology, or to, for that matter, any discipline thoroughly embedded in the theoretical tradition. It's almost like a paradigm shift where the whole methodology is revised, altered, promising new approaches to the same traditional problems. Of course, one may not agree when it is said that they are the same traditional problems. The arch light has shifted from "searching for the certainty criterion," to abandoning this search, to establishing the causal link between the "input" and the "torrential output" in the epistemic agent. There has been a shift in focus, a transfer of weight. Some even say that the transition is from normative epistemology to descriptive epistemology. But epistemology cannot be without the epistemic agent. So the issue cannot but revolve around the cognizer. This new alternative has reinforced the centrality of the cognizer in the midst of any study concerning epistemology. The cognizer needs to be observed, studied and the data thus acquired to be analysed in the light of scientific methods.

II

Going Beyond Analytical Epistemology: In Search of the Actual Knower

Analytical epistemology operated within a circumscribed framework where *knowledge* is required to pass the JTB criterion. It told us how an *ideal knower should be,* more precisely, who is an *ideal* knower? By setting normative standards of justification of beliefs and thereby defining the normative criterion that every piece of knowledge should pass, the ideal knower is forced to remain secluded from his/her individual situation/context and completely isolate his/her emotions, thereby attaining objectivity. It is only with such dissociation that one is able to attain *knowledge* and be a perfect knower. But it does not give the perspective of the *actual knower*. It offers prescription/advice for ideal cognizers rather than for cognizers with limited cognitive resources. Analytical epistemology's inadequacy perhaps lies in the fact that it falls short of explaining how ordinary beings acquire their belief and knowledge in the course of their very ordinary existence. Within the traditional framework, rationality and justification are almost co-extensive. If one is justified, then one is assumed to be rational and if one is unjustified (lacks justification) then one is assumed to be irrational. A justified belief will entail a rational belief and an unjustified belief will entail an irrational belief. And an extension of this will be — a justified belief will result in rational acts and an unjustified belief will result in irrational acts. However, if we look into the ongoing debate on human rationality, we observe that such direct and obvious association between justification and rationality is not that evident. Human cognizers consistently fail to pass normative criterion in many contexts such as having only justified beliefs, deciding only on the logical/rational acts, etc. There is an observed gap between a perfect epistemic/rational agent and an actual epistemic/rational agent. Researches into how *people actually reason* unravel the intermingling between justification, rationality and irrationality.

The key players in this intense area of research on human rationality have been philosophers and cognitive scientists, especially the cognitive psychologists. The main difference between the philosophers and the cognitive psychologists is that the former proposes the *normative* model of reasoning; whereas the latter proposes the *descriptive* model of reasoning. Philosophers aim to study — how people *ought* to reason, whereas the psychologists aim to study — how people *actually* reason. Mainly three kinds of rationality projects evolved — *descriptive, normative and evaluative.*[7] *Descriptive project*, mostly pursued by psychologists, aims at characterizing how people *actually* reason and discovering the psychological mechanisms and processes that underlie the patterns of reasoning that are observed. *Normative project,* on the other hand, aims at showing how people *should* reason. Its main objective is to discover the rules or principles of correct reasoning. It sets standards against which the quality of human reasoning can be evaluated, and can be seen as setting the parameters of human rationality.

7. R. Samuels, S. Stich, L. Faucher, "Reason and Rationality", in *Handbook of Epistemology*, eds., I. Niiniluoto, M. Sintonen, and J. Wolensk, Dordrecht: Kluwer Academic Publisher, 2004.

Lastly, *evaluative project* aims at determining the extent to which human reasoning accords with appropriate normative standards. It measures the extent to which human reasoning meets some given criterion of good reasoning.

The *philosophers* mainly argue that human beings are perfect rational beings and human irrationality is a conceptual impossibility. They argue that a competence model of actual human performance is co-extensive with the normative model. According to them, there is no gap between the descriptive and the normative models of human reasoning, and it is impossible to demonstrate systematic human irrationality. The actual observed performance is not strikingly different from the normative dictates. There are four main philosophical arguments forwarded in favour of this stand — *the argument from charity, the argument from evolution, the argument from reflective equilibrium*, and *the argument from content.*

The *argument from charity* (Quine, Dennett, Stich, Stein) is derived from the *principle of charity* as developed by Quine and Davidson and is used as a guide for interpreting human reasoning competence. Quine, in his book, *Word and Object*,[8] developed the principle of charity as a principle for guiding language communication, understanding, and translation. According to him, in matters of communication with an individual, belief ascription, or translation of some other unknown language, we have to believe that the individual is rational. In cases of interpreting utterances of another person, the utterances are seen as reflections of one's beliefs and that rationality is a necessary precondition for a person to have beliefs. If we do not start with this necessary presupposition and take people as irrational then we cannot communicate or interpret what others are saying. Different philosophers have

---

8. Quine, *Word and Object*, Cambridge, MA: MIT Press; 1960.

argued for applying the principle of charity to the interpretation of the principles of reasoning embodied in our reasoning competence,[9] and the idea is to think of interpreting a principle in our reasoning competence as analogous to translating an utterance. Proper interpretation of an individual's behaviour requires that the rationality of the individual be presupposed; for if the individual is not rational then it becomes impossible to attribute beliefs, thoughts, and judgements to that individual. If we are to characterize a person's reasoning competence, we have to assume that the principles that are being used are also rational. There are two versions of the argument from charity — the strong version of the principle of charity and the weak version of the principle of charity. The weak version of the principle of charity applied to reasoning is that unless there is strong empirical evidence against human rationality, people should be treated as rational. However, according to the strong version of the principle of charity applied to reasoning, people should never be interpreted as irrational. Any deviation from the standards of rationality should be treated as performance errors. Daniel Dennett[10] sanctioned a strong version of the principle of charity. He argued that when we approach another intentional system we could accurately predict its behaviour provided we make

---

9. There is a competence-performance distinction applied to reasoning to distinguish between inferences made in accordance with a person's reasoning competence and those resulting from various interfering factors that constitute performance errors. The rationality thesis holds that the divergence from the normative principles of reasoning constitutes a performance error and not competence error. Our reasoning competence, that is, our underlying capacity for reasoning, can be characterized by the rules that match the normative principles of reasoning.

10. D.C. Dennett, *The Intentional Stance*, Cambridge, MA: MIT Press; 1987.

three strong assumptions, which nearly give us a perfect rationality. His three assumptions are:

1. the intentional stance requires the assumption that a system's beliefs are those that it ought to have, given its perceptual capacities, its epistemic needs, and its biography. These beliefs ought to be true and relevant to one's life;
2. the system has those desires that it ought to have, given its biological make-up and its environmental set-up;
3. the behaviour of the system will consist of those acts that would be rational for the agent with the set of beliefs and desires that the agent has.

These assumptions suggest an almost perfect rationality for an individual possessing almost a perfect epistemic rationality (assumption i); having rational desires (assumption ii); and the processes of practical rationality are completely optimal (assumption iii). According to Dennett, it is necessary to make these assumptions, for otherwise we would fail to ascribe any belief to any individual. The empirical evidence from the experiments cannot be interpreted as evidence of human irrationality; they only indicate performance errors. Of course, the strong version of the principle of charity is not without criticisms.

Christopher Cherniak[11, 12] argues against this strong version of the principle of charity and tries to see just how rational humans need to be in order to have beliefs. He proposes *minimal rationality*, that is, the fact that sometimes the agent

11. C. Cherniak, *Minimal Rationality*, Cambridge, MA: MIT Press; 1986.
12. C. Cherniak, "Minimal Rationality", *Mind*, 90 (1981), pp. 161-83.

reasons in accordance with some of the norms of reasoning is a sufficient condition for the attribution of beliefs to an agent and for an agent to qualify for psychological explanations. According to him, if an agent has beliefs, then the agent must be interpreted as being minimally rational. If we consider the case of inferences, then there are some inferences which are harder than the others. For an individual there can be a class of inference which are so long and complex, or difficult, that they will be beyond the person's ability altogether. The presence of such limits is explained by what Cherniak calls the "finitary predicament." These finitary predicaments are like limited amount of time, memory, computational resources, etc.

Another philosophical argument forwarded in favour of the rationality thesis is from the standpoint of evolution, referred to as the *argument from evolution*. The perfect rationality assumption (as proposed by Dennett) supported by the strong version of the principle of charity is intertwined with the argument that evolution that guarantees human rationality. Dennett says that —

> The first answer to the question of why the intentional strategy works is that evolution has designed human beings to be rational, to believe what they ought to believe and want what they ought to want. The fact that we are products of a long and demanding evolutionary process guarantees that using the intentional strategy on us is a safe bet.[13]

Stephen Stich[14] argues against the view that evolution is necessarily truth preserving. It does not guarantee perfect epistemic rationality, nor does it guarantee that optimum

---

13. Dennett, op. cit., p. 33.
14. S.P. Stich, *Fragmentation of Reason: Preface to a Pragmatic Theory of Cognitive Evolution*, Cambridge, MA: MIT Press, 1993.

standards of practical rationality will be attained. Many have argued that evolution only guarantees that humans are genetic fitness optimizers in their local environment. It does not guarantee the kind of truth or utility maximizers that normative theories would demand. Stich showed how evolution might fail to select out accurate epistemic mechanisms when it proves costly in terms of organismic resources, like memory, energy, attention, etc.

However, David Papineau in his book, *Roots of Reason,*[15] gives arguments in favour of human ability to form accurate beliefs from an evolutionary standpoint. He says that —

> My explanation of human theoretical rationality . . . piggy-backs on the evolution of cognitive abilities for "understanding the mind" and for means-end thinking. I shall argue that once these other abilities are in place, then nothing more is needed for humans to achieve high levels of theoretical rationality.[16]

He shows that theoretical rationality[17] is a by-product of two other intellectual abilities — *understanding the mind* and *means-end thinking*, which are evolutionarily explicable. According to Papineau, *theoretical rationality is the rationality of the beliefs one adopts and practical rationality is the rationality of the choices one subsequently makes.* He equates the theoretical rationality

---

15. D. Papineau, *The Roots of Reason : Philosophical Essays on Rationality, Evolution and Probability*, Oxford: Clarendon Press, 2003.

16. Ibid., p. 40 .

17. There is a distinction drawn between theoretical and practical reasoning. Theoretical reasoning is associated with appropriate belief formation, belief update and linkage between other beliefs without maintaining any discrepancy or contradiction. Practical rationality, on the other hand, is associated with linking appropriate beliefs to appropriate desires, and fulfilling those desires with appropriate means.

of modes of thought with their reliability-for-truth. He deems a mode of thought to be rational "to the extent that it is an effective means to the consequence of true beliefs."[18] He treats "theoretical rationality" as a consequentialist notion. Of course, Papineau is careful enough not to argue that *truth* is the only consequence of belief-forming methods. He says,

> Indeed it is hard to imagine a realistic human perspective which ignores all other dimensions of possible evaluation in favour of truth. In particular, it is hard to imagine a realistic perspective that ignores significance and frugality; . . . To equate rationality specifically with reliability-for-truth would thus seem arbitrary to privilege one dimension of theoretical rationality over others.[19]

According to him, the human errors evident in the different psychological tests are due to their using methods of belief formation that sacrifice some degrees of reliability-for-truth for further desiderata like significance and frugality.[20] And it is unreasonable to condemn human thinkers as irrational just because they use methods whose less-than-ideal reliability-for-truth is often highlighted in these experiments.

Evolutionary psychologists have argued against the standard psychological understanding and explanations of the experimental data on irrationality, like for example, the heuristic and biases method as suggested by Kahneman and Tversky.[21] They opine that our characteristic modes of thought

18. Papineau, op. cit., p. 48.
19. Ibid., p. 49.
20. Significance is understood in the sense of delivering informative beliefs on matters of concern; and frugality is understood in the sense of not using large amounts of time and other resources.
21. D. Kahneman, P. Slovic and A. Tversky, *Judgement Under Uncertainty: Heuristics and Biases*, Cambridge: Cambridge University Press, 1982.

must necessarily be well suited to the range of environments in which they were originally selected. They picture the human mind as containing a number of self-contained and hard-wired "modules" each designed for a specific cognitive task. Now since these modules have been developed by natural selection over the last five million years, we should expect them to be good at satisfying the important desiderata, not across all imaginable contexts, but specifically in the "environment of evolutionary adaptation" in the range of contexts in which they evolved by natural selection.

Papineau proposed that our theoretical rationality relies on the evolution of two cognitive abilities — *understanding the minds* and *means-end reasoning*. *Understanding the minds* is the ability of humans to attribute a wide range of mental states to each other and to use this to predict and explain behaviour (Davies and Stone;[22, 23] Carruthers and Smith[24]). There is, however, no single unanimous theory about how we "understand other minds." But what is important, according to Papineau, is that any being who has an understanding of mind must inevitably have the ability to differentiate between a true and a false belief. Experiments have been conducted to test the human capacity of identifying true and false beliefs, like the false belief task.[25] The second cognitive ability *means-*

---

22. M. Davies and T. Stone (eds.), *Mental Simulation*, Oxford: Blackwell, 1995.
23. M. Davies and T. Stone (eds.), *Folk Psychology*, Oxford: Blackwell, 1995.
24. P. Carruthers and P. Smith (eds.), *Theories of Theories of Mind*, Cambridge: Cambridge University Press, 1996.
25. In this experiment, a story is narrated to the subject: "Sally puts her sweets in the basket. While Sally is out of the room her mother puts them in the drawer." The experimenter then asks the subject — "When Sally comes back, where will Sally look for →

end reasoning, on the other hand, seems so obvious to almost all of us, where humans often figure out which possible actions are the best means to their ends that it is hardly put to question. There is hardly any biological puzzle involved in this kind of reasoning. Papineau speaks of *means-end reasoning* in terms of the development of a cognitive mechanism "that takes in representations of general causal information and processes them in such a way as to select *novel* actions — that is, roughly speaking, actions which have not previously benefited the individual organism nor its ancestors."[26] He contrasts this notion of "means-end reasoning" with older cognitive systems of stimulus-response links laid down by genetic hard-wiring and associationist conditioning. He sees this means-end reasoning as an evolutionary add-on, an extra way of adjusting the stimulus-response and that which stands alongside with the older structures. His argument can be summarized as —

> Once a person identifies the end of truth (from the person's understanding of the mind), and can figure out which strategies are the best means to this end (from means-end reasoning), then one will therewith have the ability to adopt

---

→ her sweets?" If the subject has a full-fledged understanding of the mind, then the subject will be able to answer that Sally will look in the basket. Even though the sweets are really in the drawer, subjects with an understanding of mind will know that Sally's actions are guided by *her* beliefs about the world, not by the world itself, and moreover that beliefs can represent the world as other than it is, as with Sally's belief about where the sweets are. There is now fairly clear-cut evidence that all normal human children acquire the ability to pass the false belief test between the ages of 3 and 4, but not before. Papineau, op. cit., pp. 63-64.

26. Ibid., p. 67.

> reliable belief-forming methods in pursuit of true beliefs, without any further biological evolution needed.[27]

The emphasis of *argument from evolution* was that evolution has made us rational, and has favoured inferential strategies that yield true beliefs.

The third philosophical argument for the impossibility of human irrationality referred to as the *argument from reflective equilibrium* derives from an analogy to the competence/ performance distinction in linguistics. In linguistics, normative principles derive from intuitions about grammaticality. These intuitions about grammaticality are in turn based on our linguistic competence. Thus, in linguistics, normative principles are directly indexed to linguistic competence and the two cannot be separated. The idea of reflective equilibrium argument is that human reasoning competence cannot diverge from the normative principles of reasoning because both are intimately connected with our intuitions about what constitute good reasoning; and both come from the same source. The argument in its simplest form is as follows:[28]

1. The normative principles of reasoning come from our intuitions about what constitutes good reasoning.
2. Our intuitions about what constitutes good reasoning come from our reasoning competence.
3. Therefore, the normative principles of reasoning come from our reasoning competence.
4. Therefore, reasoning competence must match the normative principles of reasoning.

---

27. Papineau, op. cit., p. 73.

28. E. Stein, *Without Good Reason: The Rationality Debate in Philosophy and Cognitive Science*, Oxford: Clarendon Press, 1996, p. 137.

L.J. Cohen[29] imported the argument from linguistics into the domain of human reasoning. He argued that the normative principles of what constitutes good reasoning comes from our intuitions; and our intuitions of what constitute good reasoning comes from our reasoning competence. Thus, the normative principles of reasoning are indexed to our reasoning competence. According to Cohen, humans are by definition rational, and no experimentation is required to demonstrate that. He relied on the concept of reflective equilibrium to justify his arguments. Reflective equilibrium is achieved by a process of constraint satisfaction whereby:

> Rules and particular inferences alike are justified by being brought into agreement with each other. A rule is amended if it yields an inference we are unwilling to accept; an inference is rejected if it violates a rule we are unwilling to amend. The process of justification is the delicate one of making mutual adjustments between rules and accepted inferences.[30]

Cohen suggests that people may make performance errors on reasoning tasks, but this should not lead us to a diagnosis of irrationality at the level of competence. Cohen, therefore, claims that psychological research on reasoning falls under four categories.[31] First, the research may reveal the conditions under which subjects are prone to genuine cognitive illusions, in which case special mechanisms must be postulated to explain them. Second, it may investigate the circumstances in which the subjects reason poorly because of mathematical or scientific

29. L.J. Cohen, "Can human irrationality be experimentally demonstrated?", *Behavioral and Brain Science*, 4 (1981), pp. 317-31.

30. N. Goodman, *Fact, Fiction and Forecast*, Cambridge, MA: Harvard University Press, 1965.

31. G. Botterill and P. Carruthers, *The Philosophy of Psychology*, Cambridge: Cambridge University Press, 1999, p. 118.

ignorance. He also says that the third and the fourth categories arise when experimenters ascribe deficient reasoning subjects because the former apply the relevant normative criteria in an inappropriate manner; or apply the inappropriate normative criteria respectively.

All the above three philosophical arguments do portray the perspective of the philosophers in the rationality debate. The arguments in their respective style do imply that human rationality is largely innate and human irrationality is a myth. For some, it is futile to demonstrate human irrationality through experiments, for all human reasoning frailties can be accommodated as performance errors. Their attitude reflects firmness in relying on normative standards of rationality. Before winding up with the philosopher's account, let us briefly run through the fourth argument — *argument from content.*[32] Though this is one of the philosophical arguments, cognitive psychologists have also tried to defend better performance of humans in reasoning tasks with meaningful and real-life content. According to Carruthers and Botterill, this argument imposes some constraints on possible degrees of irrationality. They say,

> This argument claims, firstly, that what it is for an intentional state to be the intentional state that it is, involves it having a specific content; and secondly, that in the case of beliefs, the content which they have depends, in part, upon their inferential role.[33]

For example, someone sincerely assents to the statement "P*Q." Here "*" is taken as a dummy sign, subject to subsequent interpretation. Now, the person also believes that P, but does

---

32. Botterill, op. cit., p. 114.

33. Ibid., p. 114.

not assent to the statement "Q" and does not believe that "Q," though still assenting to "P*Q." Here, if the thinker expressed any belief in assenting to "P*Q," it cannot have the belief that *if P then Q*. A person, who believes *P* and fails to draw the inference to *Q*, cannot believe *if P then Q*, because it is constitutive of having such a belief that one should infer the truth of the consequent from the truth of the antecedent. In other words, believing in *if P then Q* cannot imply that the thinker assents to P and believes in P, but does not assent to Q and does not believe in Q. In an implication relationship, believing in the antecedent would also lead in believing its consequent. Theorists have forwarded examples of "content-constitutive" inferential claims. And in cases where there are content-constitutive inferential links, there are also corresponding limitations on possible degrees of irrationality. In a way, our belief content determines the inference that we are to draw inferences associated to that belief. So when a person draws an inference, it is to be judged in the light of its content belief. For example, Susan's belief that "Snakes are beautiful and harmless" would be judged in context of her belief about snakes — the concepts beautiful and harmless. Though many might not share the same inference about snakes as being harmless, it cannot put Susan's inference as being irrational. However, Carruthers and Botterill comment that —

> because only some inferential links are constitutive of content, the argument from content imposes no more than an ultimate backstop against the possible degrees of unreason, leaving wide tracts of irrationality open to an ordinary thinker. At least, there is nothing in this line of argument which would lead us to question the interpretation of psychological experiments on reasoning and inference.[34]

---

34. Botterill, op. cit., p. 116.

The most well-known philosophical distinction is between *rationality of belief* and *rationality of action.* The *rationality of belief* — how accurately a person's belief network represents the external world — has been variously termed as theoretical rationality, evidential rationality, or epistemic rationality. The *rationality of action* — how well a person's actions maximize the satisfaction of their desires, given their beliefs — has been variously termed practical, pragmatic, or means-end reasoning. Many cognitive psychologists have also maintained distinction between rationality close to the theoretical and practical rationality. Kahneman has distinguished "the ability to reason correctly about immediately available information" from the "decision theory rationality [that] is defined by the coherence of beliefs and preferences." Evans and Over[35] draw a distinction between *rationality*$_1$ and *rationality*$_2$. Rationality$_1$ is defined as reasoning and acting "in a way that is generally reliable and efficient for achieving one's goals." Rationality$_2$ refers to reasoning and acting "when one has a reason for what one does sanctioned by a normative theory."

III

How Rational are We in Our Beliefs

The purpose of stating these alternative standpoints is to show how in almost all arguments, appraisal of a person's decision as rational or irrational is based on whether his/her beliefs are true or false. But obviously all beliefs that we entertain are not true. False beliefs may even arise from true empirical facts, depending on how we understand and interpret a situation or phenomena. Suppose a person believes that the rising of the sun from the east and setting in the west everyday

---

35. J. Evans and D.E. Over, *Reasoning and Rationality,* UK: Psychology Press, an imprint of Erlbaum (UK) Francis & Taylor, 1996.

is ideal or perfect. Her justification for such a perfect phenomenon is different. She believes in regularity of things in a disciplinary sense, and randomness is not right for a good life. We need to have order in everything is God's will and hence God decided that the Sun must rise and set in fixed directions. Here from an empirical point of view, the set of beliefs derived may be (ir)rational and not (scientifically) true. Her distrust in randomness may trigger numerous of her other decisions and actions which may fetch more successful results than unsuccessful ones. Thus there can be such beliefs that are internally consistent and hence rational from the logical point of view. We do entertain false beliefs, rather beliefs whose justification does not reflect reasonableness and we do take decisions and act based on these beliefs. However, all such decisions and actions may not be termed as irrational. This is because decisions based on these beliefs often help us in achieving our goals. In case of practical rationality (rationality of actions), these beliefs work fine in achieving our goals. In fact, beliefs that help in attaining one's goals can hardly be called irrational, no matter how little consensus they hold with other fellow beings. We may find it hard to term them as rational beliefs. But how do we even term them as "irrational" given their success rate in fulfilling goals? In many cases, the beliefs may initially be without adequate justification. On being pursued to think about why a person X holds a belief Y, X may be compelled to justify it afterwards, searching for adequate reasons why she holds such a belief Y. Our ordinary, regular beliefs cannot always be placed under traditional epistemic scanner. Normative epistemic scrutiny of these beliefs will lead to failure and one is compelled to wonder about the rationality epistemic criterion that we are supposed to meet to qualify for "rational" agents.

Not only our beliefs, but also the outcome of experimental studies on human reasoning showed that normative prescriptions were flouted and were often absent from actual instances of reasoning. Systematic, predictable errors and biases were observed. Amongst these biases some were termed as *belief bias*, as *confirmation bias*, etc. The former is defined as a tendency to seek information that will confirm rather than falsify existing beliefs, theories, etc.; whereas the latter is a tendency (studied in deductive reasoning) for people to judge the validity of an argument on the basis of the prior believability of its conclusion. Beliefs hold a key role in interpreting a given problem and the decisions are often made under the persuasion of our existing beliefs. The question here is, to what extent does the corpus of beliefs *actually* affect our reasoning and how does the former steer our reasoning? Let us see an example of how beliefs affect our decisions.

People are asked to evaluate arguments whose conclusions are either believable or unbelievable. It is observed that the acceptance of the conclusion is strongly influenced by the believability of the content of the argument. People tend to accept invalid arguments when the conclusion is believable, and reject the same form of invalid argument when the conclusion is unbelievable. They act on their prior knowledge-base and are predisposed to a set of prior beliefs. To cite two examples:

> No addictive things are inexpensive.
> Some cigarettes are inexpensive.
> Therefore, some addictive things are not cigarettes.

This argument and argument-form is invalid (Fallacy of Illicit Minor) but people accept this invalid argument as the conclusion is seen to be believable. *People believe the fact that there are some other addictive things which are not cigarettes.* It is expected that those who believe the above conclusion should also accept the following invalid argument of the same form.

> No police dogs are vicious.
> Some highly trained dogs are vicious.
> Therefore, some police dogs are not highly trained.

In this case, though the argument is of the same invalid form its conclusion is not believable. *People believe that police dogs are (must be) trained dogs.* Hence few people accept this argument. It is not because of the form but the believability/ unbelievability factor that play the key role.

Many would like to believe that when subjects are given such tasks it is most likely that people will be able to notice the validity and invalidity of the arguments. Students who are taught the distinction between valid and invalid arguments should be able to detect them correctly. Subjects should have been able to identify both the arguments as invalid arguments as they commit the Fallacy of Illicit Minor, though they were asked to state their believability or unbelievability regarding the arguments. Majority of subjects, however, said that the conclusion of the first argument is believable and the conclusion of the second argument is unbelievable, and it did not strike them that both the arguments were invalid in their logical form. Their believability and unbelievability were influenced only by the content of the argument and how convincing were the facts in given arguments. Unless one is explicitly asked to study the validity and invalidity of the arguments, people often overlook these formal issues of an argument structure and go for the content of the argument. It shows that the content takes precedence over the form. The familiarity of given claims and our belief in them influence our decisions and actions in many ways than we are often aware of.

IV

The Interplay

Our beliefs are representations of our perception of the external world. The transition of our physiological perception to our mental perception involves interpretation. And thus we may have different beliefs of the same event. Beliefs are acquired as well as derived. Every belief corpus is like a web, no belief stands in isolation. Beliefs are products of understanding the world and act as supportive tools in our reasoning. Their role cannot be undermined and normative standards of reasoning are of little help. The point is that justification and rationality in the traditional sense do not always go hand in hand.

There can be real instances where justified beliefs do not lead to rational decisions and actions. Suppose Sumit works in the Defence Department where one ought to be very careful with classified information. His belief is: one should be faithful to people one knows, professionally or personally (without being secretive), because transparency in relationships is a way to maintain honesty. One day Sumit's colleague Rahul confides in him that he is tense as he has misplaced some classified information but is hopeful that given a few hours he will be able to locate it. He also pleads Sumit not to inform this to their boss and if he fails to find the document he will notify the boss himself. But Sumit's belief pursues him to report the incident to the boss immediately without giving Rahul any time. The boss panics and calls for immediate action against Rahul for his carelessness and suspends him until further enquiry. But the document was found by a search team within a few hours on Rahul's unorganised desk (as he has always been this messy). Sumit's apparently justified belief led him to a rash decision which may not be a rational one to many. Likewise, unjustified beliefs may not always lead to irrational actions. Preeti's unjustified belief that women should not smoke because smoking is meant for men has led to a probably rational decision and action — "not to smoke" — thereby evading the health hazards caused due to smoking. Questions may arise how we are certain whether "this" or "that" decision is rational or otherwise. But these questions remain to be difficult ones as it depends on what would ideally be "rational." For the time being, let us take the consequence of their decisions and thereby their decisions to be rational or irrational. Our task of judging each individual's decision as rational or not would have been easier only if a "rule book of correct rational decisions for every situation" was available. But having such a "guidebook" is just impossible as it will always remain incomplete and deficient.

Belief, justification, and rationality no doubt play an important part in our cognitive acts and decisions, and their interplay is also obvious. We would all want to believe that our belief corpus is sufficiently justified and rational because we derive our confidence to act and decide (wisely) from such reliable belief base. But what needs to be looked into is how the interplay between belief, justification and rationality is linked in actuality. Such concerns will lend better perspective and flexibility in understanding the ordinary beliefs of ordinary people.

7

Rationality and Reasoning with Metaphors*

Bipin Indurkhya

I

METAPHORS and analogies are often used as tools of reasoning. Sometimes they are so subtle that we are not even aware that a metaphor or analogy is at work in providing the justification for reaching a certain conclusion. At other times, they are quite explicitly evoked and stand out larger than life. Examples of both can be found in everyday reasoning as well as in academic discourse in science, philosophy, religion, etc. Lakoff and Johnson[1] have done an excellent job at demonstrating how pervasive subtle metaphors are in everyday reasoning. For instance, the conclusion that you can "save" time by taking the subway to work instead of walking derives from the "time is money" metaphor that most of us take for granted. On the other hand, when a government announces that it is declaring a "war" on the nation's budget deficit and asks the citizens to offer sacrifices and put up with discomforts, the underlying metaphor is blatantly obvious — the "sacrifice" is increased

---

* Originally published in *New Ideas in Psychology*, vol. 25, no. 1 (2007), pp. 16-36; reprinted with permission from Elsevier.

1. G.M. Johnson and Lakoff, *Metaphors We Live By*, Chicago: University of Chicago Press, 1980.

taxes and the "discomfort" is reduced services. (See also Kempton for a study of how people have different underlying metaphors for the way home heating works, and how these metaphors influence the way these people adjust their thermostats).[2] In a beautifully written essay titled "The Myth of Metaphor," Colin Turbayne points out that Descartes' "world is a machine" metaphor has worked its way so subtly into all of the Western sciences since his time that most scientists and non-scientists alike are no longer consciously aware of it.[3]

We should emphasize that the issue here is the role of metaphor in constructing or laying out an argument, and not whether the argument is valid or not. For instance, Francesco Sizzi is reported to have used the following argument against Galileo's discovery of the satellites of Jupiter:

> There are seven windows in the head — two nostrils, two eyes, two ears, and a mouth; so in the heavens there are two favourable stars, two unpropitious, two luminaries, and Mercury alone undecided and indifferent. From which and many other similar phenomena of nature, such as the seven metals, etc., which it were tedious to enumerate, we gather that the number of planets is necessarily seven.
>
> — Quoted in Newman 1956, 822.[4]

This argument clearly employs some very overt metaphors,

2. W. Kempton, "Two Theories of Home Heat Control", in D. Holland and N. Quinn (eds.), *Cultural Models in Language and Thought*, Cambridge: Cambridge University Press, 1987, pp. 222-42.

3. C.M. Turbayne, *The Myth of Metaphor*, New Haven: Yale University Press, 1962; revised edition with an appendix by R. Eberle "Models, Metaphors, and Formal Interpretations", Columbia: University of South Carolina Press, 1970.

4. Quoted in J.R. Newman, *The World of Mathematics*, vol. 2, New York: Simon & Schuster, 1956, p. 822.

though its conclusion was proven to be otherwise later. It seems that metaphors (in the rest of this essay my term "metaphors" will subsume analogies as well) are quite commonplace in discourse, and sometimes they are there not merely for convenience but are the lifeblood of the argument because only they provide the necessary justification for accepting (or not accepting) a certain conclusion.

Given how ubiquitous metaphors are, one might wonder: when is it rational to accept a conclusion derived from a metaphor? We may consider an extreme position: where the only justification that can be offered in support of a certain assertion comes from a metaphor, the assertion must necessarily be judged irrational — or, in other words, metaphors must be banned from all rational discourse. Though such a conclusion may seem rather radical, we should emphasize that the traditional accounts of rationality tend to not consider metaphorical reasoning at all.[5] However, this extreme view precludes many instances of everyday as well as scientific discourse from being characterized as rational. In fact, some historians and philosophers of science would argue that in this traditional view all science, except perhaps mathematics, would have to be deemed irrational. For metaphors have been found to play a crucial and indispensable role in the growth and development of various sciences.[6-12] It has even been argued

5. G. Harman, "Rationality", in E.E. Smith and D.N. Osherson (eds.), *Thinking: Invitation to Cognitive Science*, vol. III, Cambridge, Mass.: MIT Press, 1995, pp. 175-211.

6. R. Boyd, "Metaphor and Theory Change: What is a Metaphor for?", in A. Ortony (ed.), *Metaphor and Thought*, Cambridge: Cambridge University Press, 1979.

7. M. Gerhart, A. Russell, *Metaphoric Process: The Creation of Scientific and Religious Understanding*, Fort Worth, Texas: Texas Christian University Press, 1984.

→

that the structures of mathematics emerge from an interactive cognitive process that is similar, if not the same, to the one underlying metaphor.[13-15] Thus, it seems that the proper answer to the question must lie somewhere in between: sometimes metaphors can be used to derive rational conclusions, but sometimes not.

But then we must ask: What makes metaphors rational at some times but not at others? Addressing this question, as we will be doing in the rest of this essay, turns out to have a reciprocal effect. On one hand, considerations of rationality qualify when reasoning from metaphors is acceptable. On the other hand, certain metaphorical conclusions that seem quite reasonable — sometimes even necessary to maintain a rational world view — have strong implications for rationality itself. We will be examining both of these issues here.

This essay is organized as follows. In the next section, I

→ 8. M.B. Hesse, *Models and Analogies in Science,* Notre Dame: University of Notre Dame Press, 1966.

9. M.B. Hesse, *The Structure of Scientific Inferences,* Berkeley, CA: University of California Press, 1974.

10. M.B. Hesse, *Revolutions and Reconstructions in the Philosophy of Science,* Bloomigton: Indiana University Press, 1980.

11. R.S. Jones, *Physics as Metaphor,* Minneapolis: University of Minnesota Press, 1982.

12. T.S. Kuhn, 1979, "Metaphor in Science", in A. Ortony (ed.) *Metaphor and Thought,* Cambridge: Cambridge University Press, 1979, pp. 409–19.

13. I. Lakatos, "Proofs and Refutations: The Logic of Mathematical Discovery" in J. Worrall and E. Zahar (eds.), Cambridge: Cambridge University Press, 1976.

14. G. Lakoff and R.E. Nüñez, *Where Mathematics Come From,* New York: Basic Books, 2000.

15. S. Mac Lane, *Mathematics: Form and Function,* Berlin: Springer, 1986.

distinguish between two modes of metaphorical reasoning, which I refer to as monotonic and non-monotonic reasoning, and point out their characteristics. In Section III, I discuss monotonic reasoning with metaphors at some length, and then point out how considerations from rationality qualify it. In Section IV, I take up non-monotonic reasoning with metaphors. I present some examples to show how it forms an indispensable part of rational discourse, and then proceed to point out the implication it has for rationality. In Section V, I present an interaction-based view of cognition, and discuss how rationality might be viewed within it in a way that incorporates non-monotonic reasoning with metaphors.

II

Two Modes of Reasoning with Metaphors

Though I presented a few examples of metaphorical reasoning in the introduction, it would be useful to explicitly clarify what metaphorical reasoning means and to introduce a couple of terms before proceeding further. In metaphorical reasoning, two distinct domains are involved, and the implicative complex (or the knowledge) of one domain is used to reason (or to make an assertion) about the other domain. The first domain is referred to as the *source* and the second domain, the one being reasoned about, as the *target*. In the examples I presented in the introduction, time, reducing the budget deficit, the world, and the Solar system are the targets of their respective metaphors, and money, waging war, a machine, and the human head are their respective sources.

Note that I did not say anything about the *similarities* between the source and the target: any similarities between the two may or may not play a role in the reasoning process, as we will see shortly.

An instance of metaphorical reasoning usually falls in one

of two classes. One class consists of what I have called *similarity-based metaphors* elsewhere.[16] Here, some similarities are noted between the source and the target, and from these known similarities, further similarities are stipulated. It is these stipulated similarities that allow the implicative complex of the source to generate assertions about the target. I refer to this kind of reasoning as *monotonic* because it extends the knowledge of the target monotonically — no conclusion that could be derived from the knowledge of the target alone is ever invalidated by the metaphor, but new conclusions are added.

An example discussed by Gentner and Jeziorski[17] illustrates how monotonic reasoning with metaphor works. They point out how Carnot used an analogy from fluid flow to heat flow to hypothesize that the rate of heat flow might be proportional to the temperature difference just as the rate of fluid flow is proportional to the difference in levels. This conclusion was based on other known similarities between the heat flow and the fluid flow; for instance that fluids flow from higher levels to lower levels and that heat "flows" from a higher temperature to a lower temperature.

In all instances of monotonic reasoning with metaphors, known similarities between the source and the target play a crucial role in the reasoning. They are seen to "justify" or provide a rationale for the assertion derived from the metaphor.

The other kind of metaphor, which I have referred to as *similarity-creating metaphor* in Indurkhya,[18] is *not* based on known

16. B. Indurkhya, *Metaphor and Cognition*, Dordrecht: Kluwer Academic Publishers, 1992.

17. D. Gentner, M. Jeziorski, 1989, "Historical Shifts in the Use of Analogy in Science", in B. Gholson et al. (eds.), *The Psychology of Science: Contributions to Metascience*, London: Cambridge University Press, 1989, pp. 296-325.

18. B. Indurkhya, *Metaphor and Cognition.*

similarities between the source and the target. In fact, in most interesting examples of such metaphors, there are no known similarities between the source and the target. However, after the metaphor is applied, if it is successful, it makes the target look similar to the source. In other words, the similarities are *created* by the metaphor. I refer to the reasoning with similarity-creating metaphors as *non-monotonic* because it typically works by causing a radical shift in the way the target is conceptualized, and the inferences that are generated by the metaphor may be incompatible with the implicative complex of the target that existed before the metaphor.

I will forgo presenting examples of similarity-creating metaphors until Section 4, but let me just stress here that the known similarities between the source and the target — that is, the known similarities *before* the metaphor — do not play any role in non-monotonic reasoning. This is the single crucial point that distinguishes it from monotonic reasoning with metaphors.

III

Rationality and Monotonic Reasoning with Metaphors

In monotonic reasoning with metaphors, more commonly known as reasoning with similarity or reasoning by analogy, the known similarities between the source and the target play a crucial role in that they are seen to provide a rational basis for stipulating further similarities. Of course, why there is such a justification is an issue that philosophers have grappled with for a long time without success. It is interesting to point out here that this problem is closely related to the Humean problem of the justification of induction.[19] Indeed, some of the theories that seek to provide justification for reasoning

19. D. Hume, *A Treatise on Human Nature*, 1739, Harmondsworth: Penguin, 1987.

with similarities are corollaries of more comprehensive theories of induction;[20-22] and some of the arguments I am going to make here parallel the ones that I have made elsewhere concerning the rationality of induction.[23]

The attempts to provide a rational foundation to reasoning with similarities can be divided into two classes: logical and empirical. To provide a logical foundation, the goal is often to show that an inference derived from a similar source is more probable than, say, a random inference.[24] Many elaborate mathematical systems have been proposed that try to demonstrate this in one of two ways. In the inductive approach, the idea is to show that known similarities between the source and the target form confirming instances of the hypothesis that the source and the target are similar.[25-29] In the generalization approach, the hypothesis that known similarities imply inferred similarities is analysed in terms of the scopes

---

20. R. Carnap, *Logical Foundations of Probability*, The University of Chicago Press, 2nd edn., 1962.
21. R. Harrod, *Foundations of Inductive Logic*, MacMillan, 1956 (republished 1974).
22. G.H. von Wright, *The Logical Problem of Induction* (2nd rev. edn.), Oxford: Oxford University Press, 1965.
23. B. Indurkhya, "Some Remarks on the Rationality of Induction", *Synthese* 85 (1990), 95–114.
24. M.B. Hesse, *Models and Analogies in Science*, pp. 101-29.
25. R. Harrod, op. cit., pp. 123-27.
26. R. Carnap, op. cit., pp. 569-70.
27. P. Achinstein, "Variety and Analogy in Confirmation Theory", *Philosophy of Science* 30 (1963), 207–27.
28. R. Carnap, "Variety, Analogy, and Periodicity in Inductive Logic", *Philosophy of Science* 30 (1963), 222–27.
29. M.B. Hesse, "Analogy and Confirmation Theory", *Philosophy of Science* 31 (1964), 319–27.

of the antecedent and consequent, and probabilities are assigned to the generalization in direct proportion to the scope of its antecedent and in inverse proportion to the scope of its consequent.[30-31] This has the effect that a generalization inferring fewer similarities from more known similarities comes out as more probable than the one inferring more similarities from fewer known similarities.

Both these approaches are problematic, though.[32] The induction approach can be used to assign high probabilities to two contradictory conclusions made from two different sources that share the same set of similarities with the target. The generalization approach fails because it makes the existence of the source superfluous — the generalization probabilities are assigned entirely based on the scopes of the antecedents and consequents.

To provide an empirical foundation for reasoning with similarities, the usual approach is to demonstrate the pragmatic success of the reasoning process both in psychological experiments and in real-world problem-solving situations.[33-36]

---

30. J.M. Keynes, *A Treatise on Probability,* London: MacMillan and Co., 1921, pp. 222-32.
31. G.H. von Wright, op. cit., pp. 134-36.
32. B. Indurkhya, *Metaphor and Cognition,* pp. 323-27.
33. C.A. Clement, D. Gentner "Systematicity as a Selectional Constraint in Analogical Mapping", *Cognitive Science* 15 (1991), 89-132.
34. D. Gentner, "The Mechanisms of Analogical Learning", in S. Vosniadou and A. Ortony (eds.), *Similarity and Analogical Reasoning,* London: Cambridge University Press, 1989, pp. 199-241.
35. M.L. Gick, K.J. Holyoak, "Analogical Problem Solving", *Cognitive Psychology* 12 (1980), 306–55.
36. M.L. Gick, K.J. Holyoak, "Schema Induction and Analogical Transfer", *Cognitive Psychology* 15 (1983), 1-38.

But each of these approaches is also fraught with controversy.[37] The psychological experiments all satisfy the precondition that the "right" source — that is, a source that can successfully lead to the solution of the problem — is known to the experimenter. In other words, what we can conclude from these experiments is that reasoning from a source that is similar to the target can be effective provided that the "right" source is given. For real-world problems, when no such oracle is available for consulting, it is far from clear whether reasoning with similarities brings about any pragmatic success.

Most efforts to analyse real-world problem-solving situations take full benefit of hindsight, and often end up committing the fallacy of mixing up before-metaphor situation with after-metaphor situation.[38] When a particular metaphor is successful in solving a real-world problem, it always ends up making the post-metaphor target seem similar to the source — we will see some examples of this in the next section — but that does not necessarily imply that similarities were there before the metaphor, or that similarities were the reason for choosing the source in the first place.

To appreciate the futility of seeking a new or distinctive rational basis for reasoning from similarities alone, one must look at some examples where stipulating further similarities from known similarities results in an unwarranted conclusion. We come across such examples everyday in the media, in political rhetoric, and in advertisements. We are not always

37. B. Indurkhya, "Predictive Analogy and Cognition", in K.P. Jantke (ed.), *Analogical and Inductive Inference*, Lecture Notes in Artificial Intelligence 642, Berlin: Springer-Verlag, (1992b), pp. 214-31.

38. D.A. Schön, *Displacement of Concepts*, New York: Humanities Press, 1963. T.S. Kuhn, 1979, "Metaphor in Science", in A. Ortony (ed.) *Metaphor and Thought*, Cambridge: Cambridge University Press, 409–19.

aware of them. Sometimes, we already believe in the conclusion that is being derived from the similarities (so that no justification is really needed, as far as we are concerned). Sometimes, the analogy is so subtle that we find them psychologically compelling. It is only after we subject each argument from similarities to a careful logical analysis,[39] we become aware of the abuses of analogy.

Elsewhere I have presented some examples of what I called the "dark side" of reasoning from similarities, because proponents of analogical reasoning conveniently choose to ignore them.[40] For instance, Robbins[41] pointed out similarities between the 1991 debate in the US Congress to authorize the President to use force against Iraq, and the 1939 debate when the US was considering whether to intervene in Europe. This article appeared when the multinational forces led by the U.S. stood poised to attack Iraq in a bid to liberate Kuwait, and there was a world wide debate on whether military force should be used. The author claimed, "The similarity is more than superficial," but no single explanation was offered as to how. Perhaps the author felt that the similarities between the arguments used in the Congress alone warranted the conclusion that the 1991 situation in the Middle East is similar to the 1939 situation in Europe in other respects as well.

This is where Weitzenfeld's approach to reasoning from similarities is most enlightening.[42] He pointed out that there

39. G.N. Nosich, *Reasons and Arguments*, Belmont, CA: Wadsworth Publishing Company, 1982.

40. B. Indurkhya, "Predictive Analogy and Cognition".

41. J. Robbins, "Echoes of 1939 on Capitol Hill", *Wall Street Journal*, 14 January 1991, A12.

42. J.S. Weitzenfeld, "Valid Reasoning by Analogy", *Philosophy of Science* 51 (1984), 137-49.

is always a second-order generalization, namely that the similarities in the known respects *determine* similarities in some unknown respects, that underlies any argument from analogy. To use the terminology of Stephen Toulmin, it is the second-order generalization that acts as the *warrant* of the argument from similarities.[43]

There are two major consequences of Weitzenfeld's approach. One is that an argument based on similarities, and similarities alone, is not justified at all in any sense. The other is that the justification for an argument from analogy depends directly on the degree of rationality one assigns to the implicit second-order generalization. For example, to hold that similarity in nationality of two persons determines similarity in their native language seems quite reasonable, given the background general knowledge about languages people speak in different nations. But to maintain that similarities in make, model, and year of two automobiles determine similarity in their colour seems groundless (unless there is some background knowledge connecting the make, model and year of an automobile with the colour).

Thus, the problem of finding rationality in reasoning from similarity turns out to be no different from the problem of finding rationality of any hypothesis. Similarities alone do not add the least bit of justification to an argument that cannot stand on its own — they act only as a cognitive smokescreen to hide the background assumption (the second-order generalization) so that the proponent of the argument does not have to make an honest effort at justifying or defending it. This background assumption is the only warrant for an argument based on similarities, and ought to be explicitly justified, if it is not obvious already. Failing to do so turns

43. S.E. Toulmin, *The Uses of Argument*, Cambridge: Cambridge University Press, 1964.

similarity-based reasoning into a cognitive blind spot that can delude us into accepting certain conclusions as rational, when, in fact, they are not justified in any sense at all.

IV

Non-monotonic Reasoning with Metaphors

Non-monotonic reasoning with metaphors is characterized by metaphors that are not based on existing similarities between the source and the target, rather, if successful, they *create* the similarities between the two. I am going to introduce this kind of reasoning with some examples in creative problem-solving, jurisprudence and the history of science.

IN CREATIVE PROBLEM-SOLVING

Perhaps the most classic example of non-monotonic reasoning with metaphors is a case discussed by Schön.[44] A product-development team was given the task of improving the performance of synthetic fibre paintbrushes, which left the painted surface gloppy and uneven, as opposed to the smooth and continuous finish of natural fibre paintbrushes. Initially, their understanding of the painting process, and the role of paintbrush in it, was based on seeing it as a smearing process: the paint sticks to the fibres when the brush is dipped in it, and when the brush is moved back and forth on the surface, some of this paint sticks to the surface. From this perspective, when the product development team compared the performances of the natural fibre brush and the synthetic fibre brush, they could find no differences. Yet the appearance of the painted surface was different.

The breakthrough came when a member of the team had a flash of insight suggesting that the painting process be viewed as pumping. The concept of painting, as well as the role of a

44. D.A. Schön, op. cit., pp. 74-76.

paintbrush in it, was completely transformed. Now the paint is sucked up in the space between the fibres by capillary action as the brush is dipped in the can of paint, and, when the brush is moved back and forth on the surface, the increased pressure pumps out the paint through the space between the fibres, generating a thin spray at the place where the fibres bend away from the surface. When the performances of the natural fibre paintbrush and the synthetic fibre paintbrush were compared from this perspective, it was found that the natural fibres formed a smooth curve when the paintbrush was pressed against a hard surface like a wall, while the synthetic fibres bent at a sharp angle. This observation led to a number of innovative suggestions for fabricating gradual-bending synthetic fibres, some of which resulted in bristles that curved smoothly; the paintbrush made from these fibres did, indeed, produce a smooth painted surface.

There are a few important points that must be noted in this example. First of all, the two different ways of viewing the painting process structure it in radically different ways. The very causal structure that determines how the paint sticks to the brush after it is dipped in the can, and how it is coated on the surface when the brush is moved back and forth across it, is different in each case. Moreover, the primitives underlying the structures are different. For example, the space between the fibres plays no role in the structure of *painting-as-smearing* model but is a key factor in *painting-as-pumping* model.

There are two other crucial points. The flash of insight about painting as pumping was not triggered by some existing similarities between the painting process and the pumping process. This is why the identification was shocking at first — a characteristic that accompanies all creative but insightful metaphors. Secondly, the similarities between painting and pumping were *a result* of the metaphor. That is, after a different ontology for painting was created, *then* there were similarities between painting and pumping. Schön was most vehement about both these points.[45]

There are several other such examples discussed in the literature on creative problem-solving.[46-47] Gordon, for instance, proposed a problem-solving mechanism called "making the familiar strange," to be used when the familiar perspective on the problem does not help in reaching a solution.[48] The idea here is to view the familiar problem (after having unsuccessfully tried to solve it from the conventional perspective and by familiar techniques) "strangely" by juxtaposing it with strange domains. The juxtapositions initially seem quite crazy, but then one of them might lead to an illuminating insight that can prove to be the key to the solution. This mechanism is, in fact, quite the opposite of reasoning from similarities, which looks at the most similar source to import hypotheses into the problem domain. According to Gordon's view — a view that is supported by many case studies provided there — any attempt to reason from a similar source would only lead to mundane and not particularly insightful hypotheses. This is because a similar domain is not going to change the perspective on the problem domain.

Thus, we see that non-monotonic reasoning with metaphors is an indispensable aid to creative problem-solving. Almost always, we develop ontologies and perspectives on

45. D.A. Schön, "Generative Metaphor: A Perspective on Problem-Setting in Social Policy", in A. Ortony (ed.), *Metaphor and Thought*, Cambridge: Cambridge University Press, 1979, pp. 154-283.

46. W.J.J. Gordon, *Synectics: The Development of Creative Capacity*, New York: Harper and Row, 1961.

47. A. Koestler, *The Act of Creation*, Hutchinsons of London, 1964; 2nd Danube edn., 1976.

48. W.J.J. Gordon, op. cit.

objects and situations in certain contexts. With repeated encounters, we become familiar with these perspectives, and they work well in similar contexts. However, when we encounter the same objects and situations in a radically different context, and we want to achieve certain ends, this becomes a problem and the familiar perspectives may no longer be adequate for solving it. Moreover, the context may be so different that no monotonic extension of the familiar perspective would do. What we need is a completely different perspective — as if we were encountering the object or the situation for the very first time. Techniques like "making the familiar strange" are really aimed at recovering our cognitive innocence, at least in part. But then any view of rationality ought to include them as well, for, otherwise, we would have no way to allow a shift from painting-as-smearing to painting-as-pumping in a rational discourse.

IN JURISPRUDENCE

Legal reasoning is fraught with reasoning from similar past cases. In common law systems, as in England, previous cases are often the only sources of law. Even in statutory law systems, as in most of the continental Europe, past cases serve as important beacons for guiding how statutory predicates should be applied to the particulars of any given case. For a large part, legal reasoning proceeds along familiar lines — familiar to lawyers, that is — where the particulars of a given case suggest certain legal concepts and categories that might be relevant to it, and the past cases provide a constellation of concepts and categories from which different arguments can be made as to how the given case ought to be decided. (I must emphasize that I am by no means implying that such arguments are always clear-cut, or that the past cases often suggest an unambiguous resolution of the given case.)

However, there are situations when the judges and the lawyers create new concepts and categories in order to apply a precedent to the facts of a given case, or to distinguish between them.

One such case concerns a concert violinist, Ernest Drucker, who was employed by the Metropolitan Opera (Met).[49] Drucker maintained a studio at home where he practised regularly, and claimed a home-office deduction on his income tax returns. Internal Revenue Service (IRS) disallowed the home-office deduction, noting that under the tax statute, home-office deductions can be claimed only if the home-office is the "principal place of business" for the taxpayer, and arguing that Drucker's principal place of business was the concert hall and not his home studio.

Drucker petitioned the Tax Court, where the majority opinion sided with the IRS. They argued that Drucker was an employee of the Met, and consequently, was in the business of being a concert musician for the Met. Therefore, his principal place of business was the "focal point" of his activities as a musician for the Met, namely the concert hall. I must point out here that the Tax Courts had used the "focal point" test consistently in the past to determine the taxpayer's principal place of business. According to this test the principal place of business is the place where the most visible part of the taxpayer's business activity takes place: for example the principal place of business for a teacher is the classroom, and not her office where she may prepare lecture notes and grade papers, even though she may spend significantly more time there than in the classroom.

On appeal, the appellate courts reversed this decision. However, the majority opinion did not seem to want to disturb the focal point test, nor did they seem to wish to deny that

49. *Drucker* vs *C.I.R.*, 715 f, 2d 67, 1983.

Drucker was in the business of being an employee of the Met. So they made a distinction between the principal place of business of an employer and the principal place of business of an employee, and argued that the two do not always have to coincide. The majority opinion held that the focal point of *Drucker's* business was his home practice, and, consequently, his principal place of business was his home studio, though the principal place of business of his employer, the Met, was the concert hall.

The important thing to notice here is that this distinction between the principal place of business of an employee and the principal place of business of an employer was made here for the first time, it seems, solely to justify the decision for Drucker without overruling the precedents. This distinction created a new ontology, on which basis the facts of Drucker's situation and tax responsibilities could be distinguished from the other precedents.

In another example, the Tax Court allowed home-office tax deduction for an anaesthesiologist, Nader Soliman, who maintained an office at home from where he did administrative work like billing his patients. (*Soliman* v. *C.I.R.*, 94 T.C. 20). (This decision was later affirmed by appellate courts, 935 F.2d 52, 4th Cir. 1991, but reversed by the Supreme Court, 113 S.Ct. 701.) However, there was a preceding case where an emergency care physician, Stanley Pomarantz, was denied a home-office deduction in similar circumstances. (*Pomarantz* v. *Commissioner*, 867 F.2d 495, 9th Cir. 1988.) In order to distinguish the Soliman case from the Pomarantz case, the majority opinion on the Tax Court described the similarities between the two cases as "superficial" and argued that the facts of the two cases are quite different because Pomarantz spent an "insubstantial" amount of time at his home-office (8 to 15 per cent), whereas Soliman spent a "substantial" amount of time at his home-office (slightly over 30 per cent.) Here a new ontology was created by introducing the category "substantial" under which the facts of the two cases were distinguished.

Thus, we see that legal reasoning sometimes requires the creation of new categories, or a change of ontology, especially when unconventional and creative arguments are being made. The created categories can make the facts of two different cases seem similar, or they can distinguish between cases that would otherwise seem similar.

IN THE HISTORY OF SCIENCE

Historians of science have pointed out that in many instances the development of science was crucially affected by certain analogies or models.[50-56] While some of these analogies were guided by the then-known similarities between the source and the target, as in the example of Carnot's use of the fluid flow analogy to reason about the heat flow that was mentioned earlier, in many cases the source was applied to the target even though there were no known similarities

50. Canguilhem, G., "The Role of Analogies and Models in Biological Discovery", in, A.C. Crombie (ed.), *Scientific Change,* New York: Basic Books, 1963, pp. 507-20.
51. H.E. Gruber, "Darwin's 'Tree of Nature' and Other Images of Wide Scope", in, J. Wechsler (ed.) *On Aesthetics in Science,* Cambridge, MA: MIT Press, 1978, pp. 121-40.
52. M.B. Hesse, *Models and Analogies in Science.*
53. A. Koestler, op. cit.
54. W. Libby, "Scientific Imagination", *Scientific Monthly,* XV (1922), 263-70.
55. A.I. Miller, "Visualization Lost and Regained: The Genesis of the Quantum Theory in the Period 1913-27", in J. Wechsler (ed.), *On Aesthetics in Science,* Cambridge, MA.: MIT Press, 1978, 73-102.
56. D. Rothbart, "The Semantics of Metaphor and the Structure of Science" *Philosophy of Science,* 51 (1984), 595-615.

between the two phenomena at that time. These latter kinds of analogies, which have also been referred to as "pre-theoretic models" or "material models," usually work as follows. There is a target phenomenon that is poorly understood, and a source that is more concrete and well understood. But there is no good connection or transference between the two phenomena that would form the basis for the source to be meaningfully applied to the target. There is only some vague idea in the mind of the scientist. Sometimes, there is not even a vague idea but an emotional drive that keeps the scientist seeking some connection. However, once the connection is found, if it is found, the model starts to become entrenched. If we compare the states of the source and the target before and after the model became entrenched, the creation of similarities can be seen quite starkly. The source is seen to create an ontology for the target phenomenon, where none existed before.

I present two examples here to illustrate these points. Gruber,[57] on analysing Darwin's notebooks that he kept while working on his celebrated theory of evolution, noted that the image of an irregularly branching tree kept on recurring in his thoughts, and may have served as the primal metaphor for his natural selection principle. Gruber's account cites several excerpts from Darwin's notebooks to show that it was not some existing similarities between the tree image and whatever was known about the evolution at that time that kept Darwin searching for what other similarities might be found. On the contrary, it was an emotional drive that kept up his intellectual commitment to articulating an account of evolution that matched the tree image. Darwin was not searching for similarities, he was formulating them. This process was actually

57. H.E. Gruber, op. cit.

fraught with several problems that appeared to invalidate the model, but Darwin's intellectual commitment to the idea kept him from throwing away the image altogether. Instead, he improvized and sought different ways to render the image meaningful.

In another example, Solomon[58] points to Wegener's use of the iceberg analogy to suggest the theory of continental drift.[59] Solomon notes that the theory of continental drift was not widely accepted by the geologists until the mid-1960s — after the discovery of seafloor spreading, which led to the development of the modern theory of plate tectonics — and was considered quite revolutionary when Wegener first proposed it in 1915. Here again, the analogy was not based on the known similarities between the continents and the polar icebergs. In fact, Solomon explicitly notes: "[T]he situations of icebergs and continents [were] not shown to be sufficiently similar to support the conclusions drawn."[60] I should also point out that Solomon attributes Wegener's strong attraction to the drift theory partly to the fact that he was not trained as a geologist, and so he was not indoctrinated with the contractionist theory — the leading geophysical theory of continents at that time. The reader should be able to see a connection here with Gordon's *making the familiar strange* and Schön's story of paintbrush-as-a-pump metaphor. In order to conceptualize a phenomenon differently, and to see it from a different ontology, it is necessary that one be not attached too strongly to the conventional or familiar way of looking at it.

58. M. Solomon, "Scientific Rationality and Human Reasoning", *Philosophy of Science*, 59 (1992), 439-55.

59. A. Wegener, *The Origin of Continents and Oceans*, 1915; 4th edn. (1966), tr. J. Biram, Dover, London.

60. M. Solomon, op. cit., p. 447.

Thus, we see that non-monotonic reasoning with metaphors — reasoning that is not triggered by some existing similarities between the source and the target, but is constitutive of them — plays a significant role in the development of science, and must not be excluded from rational discourse.

IMPLICATIONS FOR RATIONALITY

The preceding examples suggest that metaphors can radically alter the ontology of the target, thereby creating a new perspective on it that includes information not contained in the old representation. Moreover, this new information often contains the key to resolving a problematic situation that seems impossible from the familiar perspective. This role of metaphors, though it is essentially non-algorithmic and heuristic-driven, should not be excluded from an account of rationality, because, otherwise, one would have to admit irrationality as a key ingredient of many major areas of human thought and discourse.

The whole problem boils down to incorporating an ontology-changing move into rationality. This move should allow introducing new concepts and categories to describe an object or a situation. Moreover, the new concepts and categories are not always added to the old concepts and categories, but may replace them — in the context of the current problem, if not always. Thus, any account of rationality must

1. explicitly take into account the ontologies of representations,
2. contain mechanisms for changing ontologies, and
3. allow different representations of the same object or phenomenon to coexist, and be able to weigh their relative merits in any given context.

Allowing an ontology-changing move into rational discourse is not as simple as it may appear at first. Though a rational cognitive agent may initiate a change of ontology, the new ontology may not always lead to the desired effect. The following examples illustrate this problem:

1. Schön noted that the group working on the synthetic-fibre paintbrush problem also considered "painting as masking a surface" metaphor, with a corresponding change of ontology, but it led to no useful insight.[61]
2. Yolanda Baie, a food stand operator and owner, petitioned for having the kitchen of her house, where she prepared food sold at the food stand, qualify as a "home-office" for the purpose of tax deduction (see *Baie* v. *C.I.R.*, 74 T.C. 105). One of her arguments was that her kitchen was a "manufacturing facility" for her business. The judges, while finding the argument "ingenious and appealing," ruled it "insufficient" nevertheless.
3. John Casti[62] comments on the fate of Immanuel Velikovsky's theory as outlined in *Worlds in Collision,* which hypothesized Earth's encounters with a large comet expelled from Jupiter and provided explanations for many biblical events. Velikovsky's theory proposes a revolutionary new ontology for understanding the Solar system. But it has not been accepted by the scientific community.

Thus, the problem is to somehow allow creative changes of ontology, but to rule out arbitrary changes of ontology; or, at least, to constrain which inferences are rationally acceptable

61. D.A. Schön, op. cit., p. 259.

62. J.L. Casti, *Paradigms Lost: Images of Man in the Mirror of Science,* New York: William Morrow & Co. Inc., 1989; see pp. 7-10.

from the changed ontology.[63] In the next section, I outline a view of cognition that addresses this problem.

V

Cognition and Rationality

The problem of rationality mentioned at the end of the preceding section is remarkably parallel to a problem of creative metaphors: namely, "how can some metaphors create new and useful perspectives on the target, whereas many others simply do not work?; It also parallels a problem of cognition: how can a cognitive agent view an object (or an event, or a phenomenon) in many different ways, but not in any arbitrary way? As I have discussed both these latter problems at length elsewhere[64] by proposing a view of cognition as an interaction and articulating a theory of metaphor within it, I will attempt here to address the problem of rationality within the same framework.

AN INTERACTION-BASED VIEW OF COGNITION

It will be useful at first to lay down the rudiments of the interaction view of cognition, which has been influenced by the works of Ernst Cassirer,[65] Nelson Goodman,[66] and Jean Piaget.[67-69] Cognition is seen as an interaction between the

63. S. Palmer, 1992, *Human Ontology and Rationality*, Avebury, Aldershot, England, 1992; see pp. 62-64.

64. B. Indurkhya, *Metaphor and Cognition.*

65. E. Cassirer, *The Philosophy of Symbolic Forms*, vols. 1-3, New Haven: Yale University Press, 1955.

66. N. Goodman, *Ways of Worldmaking*, Indianapolis: Hackett, 1978.

67. J. Piaget, *Biology and Knowledge*, Chicago: University of Chicago, 1971.

68. J. Piaget, *Genetic Epistemology*, New York: W.W. Norton, 1971.

69. J. Piaget, *The Origin of Intelligence in the Child*, New York: Penguin, 1977.

cognitive agent and the external objective world; the agent and the environment are deemed to play an equal role in determining cognitive structures, which include scientific theories.

The key features of our view are as follows. We maintain a sharp distinction between an object (or an experience, or a domain, or an environment) and its representation (or description) and argue that an agent can cognitively access an object only by interacting with it in some way. Further, we claim that in order to have purposeful interactions with the world, the cognitive agent must use some internal states or structures. We refer to these internal states as "representations."

I must emphasize here that our use of the term "representation" is somewhat different from the way it has generally been used in artificial intelligence and philosophy of mind. Our "representation" is a meta-level term that refers to the internal states of the cognitive agent. We do not wish to imply that the cognitive agent has to be consciously aware of the representations or what they mean (how they are connected to the world). We do not even aim to imply that the representations have to be symbolic, though, talking about the internal states from a meta-level, we can certainly assign symbols to them. (There is an example discussed in our earlier work[70] where representations are internal states of a simple cognitive agent. Therefore, arguments *a la* Brooks[71] do not automatically run contrary to our position.)

We distinguish between the ontology and the structure of a representation. The ontology is the set of primitives in terms of which the object is described, and the structure is the particular way in which the primitives are interrelated in the

70. B. Indurkhya, *Metaphor and Cognition*, pp. 135-51.

71. Brooks, R.A. "Intelligence without Representation", *Artificial Intelligence* 47 (1991), 139-59.

description. When the cognitive agent has internal states, the set of states constitutes the ontology of the representation, and the state transition functions make up the structure. An important observation to make here is that every structure presupposes an ontology.

When a cognitive agent interacts with an object, the result is a representation of the object that is internal to the agent. This representation is determined in part by the agent and in part by the autonomous nature of the object. The agent gets to play a role in giving an ontology to the object, a process which may be partly constrained by the agent's own biological structure. The object gets to play a role by endowing this agent-determined ontology with a structure. Thus, the structure of the object always seems autonomous (external) to the agent, although the agent can choose a different ontology, and thereby affect the structure of the object as internally represented.

To elaborate on this representation-building process, we need to make an explicit distinction between its two poles. On one end are the concepts, which are the building blocks of representation, and on the other end are the sensory stimuli, which provide the raw data out of which representations are formed. Both these ends are crucial in our view of the representation-building process, in keeping with Nelson Goodman's rephrasing of the Kantian dictum:

> Although conception without perception is merely *empty*, perception without conception is *blind* (totally inoperative).[72]

We refer to these two components of the representation-building process as concept networks and sensorimotor data sets respectively.

---

72. N. Goodman, *Ways of Worldmaking*, Indianapolis: Hackett Publishing Co., 1978; see p. 6 (emphasis original).

Concept networks are structured networks of uninterpreted symbols and relations between them. The theory of pumping, the theory of iceberg drift, geometric concepts like "triangle" and theorems relating them are all examples of concept networks, as long as no interpretation is attached to them, meaning that the theories are not applied to the world. I must hasten to add that we do not wish to suggest that these theories are devoid of any meaning for the cognitive agent. Obviously, all these theories were developed to explain certain phenomena and are, therefore, representatives of them. All we mean to say is that at the meta-level, in order to explicitly talk about the representation-building process, we can sever the meaning connection and refer to the now uninterpreted theories as concept networks.

A concept network may be genetically inherited, or it can be learned from the past interactions with the environment. The ontology of a concept network refers to the set of concepts, and the structure refers to the statements and theorems. Thus, if we write a theory in a formal system, its language is the ontology and the set of sentences of the theory forms the structure. Notice that the semantic content of a concept network manifests itself in its structure: the way the concepts are interrelated to each other reflects the structure of the phenomenon to which the concept network usually applies.

Sensorimotor data sets are structured sets of stimuli received from the environment, such as the set of images formed on the retina as a cognitive agent watches the process of painting. Sensorimotor data sets provide the raw input to the representation-building process. It is useful to apply the distinction between ontology and structure to the sensorimotor data sets also. The set of raw sensory stimuli, such as the set of retinal cell excitations, together with the set of correlations between different stimuli that the cognitive agent is capable of noticing, constitute the ontology of the sensorimotor data set. As our perceptual apparatus creates the sense data, it suggests that the ontology of the sensorimotor data set is determined by the perceptual and motor organs of the cognitive agent. However, as noted by Rosch,[73] reality, as it presents itself to us in the form of sensory stimuli, is already highly structured: certain sets of stimuli occur together, while certain others preclude one another. This specific pattern of correlations that actually exists between different stimuli makes up the structure of the sensorimotor data set. Moreover, this structure of the sensorimotor data set is determined by the mind-independent external world.

This distinction between the ontology (determined by the cognitive agent) and the structure of a sensorimotor data set (determined by the external world) is crucial. An example would be helpful to clarify it further. Suppose I am standing in front of a tree. It is the biological structure of my eyes that determines what image is formed on my retinas, what contrast I see, what colours I see, etc. (Of course, it also depends on the object in front of my eyes, the lighting conditions, etc. But different objects, or the same object under different lighting conditions, can all be considered different stimuli in the sensorimotor data set.) Similarly, as I walk towards the tree, it is the structure of my motor apparatus that determines what movements I can make. However, the effect of any such movement on changing one stimulus on my retina into another, possibly different, stimulus is determined by the structure of the external world. It is precisely in this sense that reality, while owing its form (as experienced by me) to my sensorimotor apparatus, none the less appears as an autonomous external entity to me. (This

73. E. Rosch, "Principles of Categorization", in E. Rosch and B.B. Lloyd (eds.), *Cognition and Categorization*, Hillsdale, NJ: Lawrence Erlbaum Associates, 1978, pp. 27-48; see p. 28.

example may seem somewhat simplistic because our sensory organs already act as pre-processors of sensory information, and one could argue that the structure of this pre-processed sensory information is due, at least in part, to the structure of the sensory organ. However, as argued by Indurkhya,[74] in this situation, the account of the representation-building process being articulated here can be applied to the pre-processing of sensory information itself: and in explaining the processing done even by simple sensory organs, one could posit a sensorimotor data set with an agent-driven ontology and an environment-driven structure.)

Finally, it must be emphasized that a sensorimotor data set includes only the perceptual data; in particular, we do not allow information obtained by instruments to be a part of it. The reason is that when an instrument is being used, it ultimately provides us with some perceptual data, such as the deflection of a needle, an image seen in a photograph or through the eyepiece of a telescope, or squiggly lines on graph paper, and one needs to apply a theory (that is, carry out the representation-building process) to interpret the perceptual data as the current flowing through a wire, a planet up in the sky, or the activity of certain brain cells.[75]

In building the representation of an object (or event or phenomenon), the cognitive agent applies a concept network to the object. Some of the time, perhaps most of the time, this process may be habitual — the cognitive agent may have biological or cultural preferences for applying certain concept networks to certain objects or phenomena in certain ways. Recall that all the examples of concept networks presented

74. B. Indurkhya, *Metaphor and Cognition,* pp. 179-87.

75. I. Lakatos, *Philosophical Papers:* Vol. 1: *The Methodology of Scientific Research Programmes,* ed. J. Worrall and G. Currie, Cambridge: Cambridge University Press, 1978; see pp. 12-19.

above have semantic contents, thereby suggesting that there are certain objects or phenomena in the world to which they conventionally apply. However, we can get a glimpse of the representation-building process when a concept network is applied to a new phenomenon (as in viewing painting as pumping), or when the phenomenon reveals unexpected behaviour (as when the synthetic-fibre paintbrushes left the painted surface gloppy and uneven). Such situations reveal two principal mechanisms of representation-building.

Projection, derived from the Piagetian notion of assimilation,[76-77] is used in applying a concept network to a phenomenon to which it does not normally apply. In this process, parts of the sensorimotor data set of the object are grouped together, and labelled with the concepts from the concept network. In the grouping process, the sensorimotor data set acquires an ontology as seen from the conceptual layer. However, the autonomous structure of the sensorimotor data set asserts itself by endowing the conceptual ontology of the phenomenon with an autonomous structure too. In the context of the paintbrush-as-a-pump example, this means that though the cognitive agent can group the process of painting in various ways so as to see it as a pumping process (this process will also be constrained by the existing semantic content of the "pumping" concept network), whether this grouping satisfies the predictions made from the concept network (such as whether certain design of synthetic-fibre paintbrush will result in an even painted surface) depends entirely upon the autonomous nature of external reality.

Projection is non-algorithmic and heuristic-driven. The reason behind it is that an ontology is a prerequisite for an

76. J. Piaget, *Play, Dreams and Imitation in Childhood,* New York: W.W. Norton, 1962.

77. J. Piaget, *The Origin of Intelligence in the Child*, 1977.

algorithm. For example, in seeing the paintbrush as a pump, there is no algorithmic way to arrive at the appropriate ontology. In fact, there is no algorithm to even decide if an appropriate ontology exists. This is why in creative problem-solving situations, it is impossible to determine a priori which metaphors will work and which will not. Projection crucially involves what Harold Brown has called *judgement*,[78] and may reflect the cognitive agent's cultural (both inherited and learned) background as well as past individual experiences.

When the object, owing to its external autonomous structure, reveals some unexpected behaviour — in other words, it was not anticipated by the cognitive agent based on the structure of the concept network that was projected on the object — the cognitive agent may choose to restructure the concept network so as to integrate the new behaviour. This process is referred to as *accommodation*.[79-80]

Unlike projection, accommodation can be quite algorithmic. This is because in accommodation, one stays within the same ontology, there is some specific structural statement (or set of statements) that is found to be incoherent with the environment, and what is done is to restructure the theory enough so that the incoherent structural statements are no longer included.[81-82] Of course, accommodation is not always deterministic, for there may be more than one way to exclude the incoherent structural statements.

78. H.I. Brown, *Rationality*, Routledge, London: 1988; see ch. IV.

79. J. Piaget, *Play, Dreams and Imitation in Childhood*, New York: W.W. Norton, 1962.

80. J. Piaget, *The Origin of Intelligence in the Child*, 1977.

81. J. Doyle, 1979, "A Truth Maintenance System", *Artificial Intelligence*, 12 (1979), 272.

82. J. de Kleer, "An Assumption-based TMS", *Artificial Intelligence*, 28 (1986), 127-62.

Although projection and accommodation act in consort in building representations, one can find examples where one or the other mechanism plays a predominant role. Such examples help to illustrate the representation-building process, and also the mechanisms of projection and accommodation themselves.

The system of latitudes and longitudes, and the division of day into hours, minutes and seconds are both examples of representations that are built primarily by projection. The method Imre Lakatos[83] has referred to as "monster-barring" in mathematical theorem proving can also be viewed as projection, because the mathematician is essentially redefining the concept (thereby changing the ontology) so as to exclude the "monster" examples. Note that projection is non-algorithmic and heuristic in all these cases. For instance, the discussion in Lakatos[84] reveals how various new definitions of "polyhedron" were coined in order to exclude certain troublesome solids. These new definitions, however, were always based on noticing (or, perhaps, "creating") some new attribute of the solids so that the attribute would apply to the desired polyhedra but not to the monsters. Some very novel attributes were created in the process. Considering them, it seems impossible that an algorithm can generate all possible attributes of a solid that could be used as a basis for creating different ontologies.

Here is an important point about projection: even though the cognitive agent chooses a concept network (with an ontology and a structure) and groups the sensorimotor data set, thereby giving it an ontology as seen from the conceptual layer, once a concept network is projected, the structure of

83. I. Lakatos, *Proofs and Refutations: The Logic of Mathematical Discovery*, eds., J. Worrall and E. Zahar, Cambridge: Cambridge University Press, 1976.

84. I. Lakatos, *Proofs and Refutations.*

the environment as seen through the projection is determined by the objective world. Once the lines of latitudes and longitudes are put in place, whether any two given places lie on the same latitude is no longer an arbitrary matter. Once a definition of polyhedron is accepted, it becomes an objective matter whether a certain solid is a polyhedron, and whether the numbers of its faces, edges, and vertices bear a certain relationship. Notice that once the ontology is fixed, algorithms can be devised to determine structure and check for structural coherency: once the system of latitudes and longitudes is projected, we can come up with algorithms to decide whether two places have the same latitude or not.

On the other hand, geographical or political maps, and the division of year into days are both examples of representations formed primarily by accommodation. Here the ontology is kept fixed, and the structure of the concept network is adapted to reflect the structure of the environment as seen from the given ontology. Lakatos's method of revising the conjecture in the face of counter-examples is also an example of accommodation.[85] Notice here the algorithmic nature of accommodation. Once the ontology of a geographical map is fixed, one can come up with an algorithm for taking measurements on the ground and transferring these measurements to the map (which corresponds to the concept network). In Lakatos' examples, existing attributes of the geometric solids are used to limit the scope of the conjecture — a process that is very different from that of inventing new attributes as in case of projection.

I must emphasize that in accommodation, the cognitive agent also gets to play a key role in building the representation by fixing the ontology, so it is not completely an environment-driven process. Moreover, in map-making at least, there are instances where ontology itself becomes an issue, thereby

85. I. Lakatos, *Proofs and Refutations.*

making projection more dominant in the representation building process. The debate over what constitutes a "wetland" is a case in point.[86]

Finally, I must note that the two-layered account of representation-building process presented here, in which projection and accommodation are at work between one concept network layer and sensorimotor data set layer, can be extended into a model where the process of integrating sense data into concepts proceeds through several layers of intermediate representations.[87]

VI

Rationality in the Interaction View

We can now address the problem of incorporating an ontology-changing move into rationality. Notice that in the foregoing account of cognition, the representation-building process pits the concept network, which is structured by the cognitive agent, against the sensorimotor data set, which is structured by the external environment. As these structures are independent, one may not respect the other; we will refer to this characteristic as *incoherency* (or *coherency*, when the two structures do respect each other). We claim that considerations of rationality demand that incoherency, whenever it is found, be duly acknowledged and dealt with; however, they do not constrain the manner in which incoherency is dealt with; in particular, they do not favour either projection or accommodation as a primary mechanism for resolving incoherency. I will elaborate on this remark in the rest of this section.

---

86. W.J. Mitsch and J.G. Gosselink, *Wetlands*, New York: Van Nostrand Reinhold, 1986; see pp. 15-20.

87. B. Indurkhya, *Metaphor and Cognition*, 179-87.

Let us say that a certain theory has been developed to explain a phenomenon. In our terminology, the theory is a *representation* of the phenomenon. In order to explicitly talk about different theories applying to the phenomenon, and the theory applying to other phenomena, we sever the meaning connection between the representation and the phenomenon. Now the theory corresponds to our concept networks and the phenomenon becomes a sensorimotor data set. Of course, the cognitive agent normally uses the theory to comprehend and interact with the phenomenon, so the structure of the concept network is really the structure of the phenomenon as seen from the conceptual layer.

Until some point in time, the theory made many correct predictions about the phenomenon. Then an event was found that contradicted the predictions made by the theory: the representation turned out to be incoherent. Rationality would demand that this contradiction be acknowledged; for it would certainly be quite irrational to simply ignore the available evidence and insist on the correctness of the theory. However, there are two ways in which the incoherency can be dealt with.

One alternative is to keep the same ontology, but use accommodation to alter the structure of the concept network: in other words, revise the statements of the theory so as to incorporate the troublesome observation (as well as all of the past observations that are relevant to the phenomenon). For instance, after noticing that the synthetic fibre paintbrushes do not produce as good-looking painted surface as natural fibre paintbrushes do, one could try to account for it by noticing some difference between the natural and the synthetic fibres, while still staying within the painting-as-smearing theoretical framework. Of course, any such explanation will be subject to further testing, and later observations may render it incoherent. In the synthetic-fibre paintbrush story, Schön observed that the members of the product development team noticed that the natural fibres had all split ends. Thinking that this explained the appearance of the painted surface, they tried to make synthetic fibres with split ends too, but it did not yield the expected result.

The other alternative is to use projection to alter the ontology of the phenomenon.[88-89] This can be done by noticing new attributes and relations in the phenomenon, using them to re-categorize the objects and events, noticing correlations between these new categories, and creating a theory to express these correlations. Or, one can take a different theory — a theory that was not developed in the context of this phenomenon — and try to understand the phenomenon in terms of the new theory, as happened in the paintbrush-as-pump example. This process of projecting a new theory onto a phenomenon is constrained from both ends. On one end, the existing structure of the concept network, which reflects the structure of the phenomenon for which the theory was originally developed, resists arbitrary application; on the other end, the autonomous structure of the phenomenon, which is rooted in the external world, resists arbitrary categorization. For instance, in seeing painting as pumping, one has to respect the conceptual structure of "pumping" as well as the structure inherent in the actual process of dipping the paintbrush in a can of paint and moving the paintbrush back and forth across a surface. When this process of projection succeeds, if it succeeds at all, what results is a new ontology of the phenomenon with a structure that reflects the structure of the new theory.

---

88. I. Lakatos, *Philosophical Papers,* vol. 1, pp. 16-17.

89. I. Lakatos, *Proofs and Refutations.*

The important thing to notice in projection is that each time the ontology is changed, the environment endows this ontology with a structure that is external, and hence appears objective, to the cognitive agent. And to stay within rational discourse, the cognitive agent must respect this objective structure. Though, of course, the cognitive agent may shift the ontology again, if not satisfied with the structure seen with respect to the previous ontology, but then the external world gets to structure this new ontology, and so the game goes on. Needless to say, the cognitive agent may resort to accommodation at any stage of the game by sticking to an ontology and altering the structure of the theory.

It is interesting to note here that projection gets to play a key role both when a revolutionary new theory with a radically different ontology is proposed — when new discoveries are made by projecting a concept network onto an "unsimilar" phenomenon and also when hypothetical entities are proposed in order to save an old theory in the face of contradictory evidence.[90]

Though considerations of rationality alone do not dictate any hard and fast rules as to when accommodation should be favoured over projection and vice versa, some general guidelines emerge if we examine carefully why a change in ontology becomes necessary sometimes. Cognition typically involves categorization: different sensory stimuli, made available to us by our perceptual apparatus, are grouped into concepts and categories. The world as seen from the conceptual level is considerably more simplified and structured than the one at the sensory level. This simplification is necessary to make us survive in an infinitely complex world with our finite and limited minds. However, an act of

90. H.I. Brown, "Assimilation and Discovery", *New Ideas in Psychology* 1 (1983), 89-97.

categorization invariably involves loss of information.[91] When a bunch of objects are placed in a category, their potential differences are overlooked; one might say that they are discarded by the process of cognition. Similarly, when two objects are put in different categories their potential common grounds are discarded. Though the simplified world view that results from grouping the world one way makes it easier for the cognitive agent to interact with the world, the agent is deprived of a horde of alternate world views.

The concept networks that a cognitive agent develops — over a lifetime or over generations — for various phenomena and objects in the environment, reflect the agent's goals and predispositions in its past interactions with those phenomena and objects. The simplified world views created by these concept networks may make it very efficient for the cognitive agent to go about its everyday activities; sometimes they may even make a crucial difference in whether the cognitive agent can survive in the environment. However, as the environment changes (e.g. synthetic fibres become available), and the cognitive agent's goals and predispositions also vary (e.g. the need to have synthetic-fibre paintbrushes arises), sometimes the information lost in the categorization leading up to the existing concept network becomes the key to realizing the agent's goals. In such a situation, projection becomes a crucial asset to cognition, for only it can bring about a change in the ontology and possibly recover (part of) the lost information.

All this suggests that whenever the domain of activity requires some stability in ontology, as in science and jurisprudence, accommodation is to be generally preferred. An ontology-changing move (projection) should be used sparingly — only on those infrequent occasions when every

91. M.B. Hesse, *The Structure of Scientific Inferences*, pp. 48-51.

effort has been made to address the problem within the existing ontology but without any success. For projection is cognitively expensive, requiring a mammoth effort to try to render the new theory meaningful in the context of the phenomenon, even before it can be ascertained whether the new ontology satisfies the desired goals. Yet, if a type of problem is particularly hard to solve from the familiar point of view, a rational agent should realize that it may be because some crucial information is missing from the existing conception of it, and a new ontology may be called for. Forgetting this, the agent may fall into the rut of viewing the phenomenon always from the same old perspective, thereby shunning all creativity. This is why Gordon[92] touts "making-the-familiar-strange" as a key mechanism of creative problem-solving.

On the other hand, there are certain domains where the goal is to create new perspectives — to reveal alternate worlds that are lost in the familiar perspective — for no other reason but their own sake. Conceptual and modern art are a good example of such domains.[93] In these situations, projection ends up playing a dominant role, for, obviously, it is the only mechanism that allows new perspectives to be created.

When we consider changes in ontology, another issue to ponder is if there are any criteria that identify certain ontology-changing moves rational and others irrational. There is obviously a difficulty in positing any such criterion a priori. This is because in projecting a new theory onto a phenomenon, one cannot determine beforehand what the new ontology of the phenomenon would be and what consequences it might have. It is for these reasons that the crux of creativity remains rather elusive.

---

92. W.J.J. Gordon, op. cit.

93. H.N. Fox, *Metaphors: New Projects by Contemporary Sculptors*, Washington D.C.: Smithsonian Institution Press, 1982.

It is tempting to connect rationality with the a posteriori success of the ontology-changing move. This would have the consequence that the three examples of failed ontology-changing move presented earlier under the heading "Implications for Rationality" would be dubbed irrational. Taking this position, however, puts too much of a burden on the concept of rationality; making rationality synonymous with correctness or truth. If we carefully analyse each of the examples of "Implications for Rationality" without considering whether the ontology-changing move was successful or not, it is hard to come up with any grounds to reject them: to view painting as masking a surface seems no more of an offbeat idea than to view it as pumping; Baie's claim that her kitchen was a "manufacturing facility" seems no more *ad hoc* than the court's argument that Drucker's principal place of business is not the same as his employer's; and Velikovsky's theory of the Solar system seems no more groundless than Wegener's theory of continental drift at the time each one was proposed.

Thus, rationality is not so much a characteristic of an argument that predicts success. Rather, it is an attitude of the cognitive agent that acknowledges incoherency, decides when and how ontology-changing moves are made, and displays a healthy respect towards the autonomous structure of the world.

8

Epistemology and Knowledge Representation

A Cognitive-Historical Approach to Philosophy of Science

Amitabha Gupta

HISTORY of Science reveals that epistemology and science, especially during the time of scientific breakthrough in the form of production and defence of new scientific knowledge, have a close and intimate relationship. According to some philosophers of science,[1] it would not be an exaggeration to say ". . . to an overwhelming degree the history of epistemology (and metaphysics) is the history of the philosophy of science. . . ."[2]

Science and Epistemic Norms

During the formative period of Classical Physics, epistemologists, such as Locke and Kant, took scientific

1. J. Losee, *A Historical Introduction to the Philosophy of Science*, New York: Oxford University Press 1972; W.T. Jones, *History of Modern Philosophy*, New York: Harcourt, Brace and World, New York: 1952; A. Levison, *Knowledge and Society*, Indianapolis: Pegasus, 1974.
2. F. Suppe, *The Structure of Scientific Theories*, Urbana, Ill.: University of Illinois Press, 2nd edn., 1977, p. 717.

knowledge as a paradigm of epistemic enterprise and looked upon themselves as crusaders for vindicating the cause of the new knowledge and epistemic claims that were being made by the scientists of their time, especially Newton. Their job as epistemologists consisted of describing the processes through which the paradigmatic knowledge is acquired and spelling out of the norms appropriate for this knowledge.[3]

At the turn of the twentieth century, when the modern physics was being shaped, Logical Positivists too felt the necessity of articulating a new paradigm of scientific knowledge, and accepted the goal of providing and justifying an account of the nature of scientific enterprise. In this endeavour, the Logical Positivists pursued with messianic fervour the search for the epistemic norms that would avoid the pitfalls of metaphysical traps in which science, according to them, could be enmeshed, particularly while asserting regularities of nature by invoking entities, which are not directly observable.

The Received/Standard View on Epistemology

Epistemology is an old discipline and Plato[4] took the first step of distinguishing opinion, belief and knowledge. However, the credit for first spelling out the defining characteristics of the *Received/Standard* epistemological view goes to Hintikka.[5]

---

3. J. Locke, *An Essay Concerning Human Understanding*, New York: Meridian Books, 1968 printing. Locke took his job to be that of an under-labourer with respect to Newton. I. Kant, *Metaphysical Foundations of Modern Science* and *Critique of Pure Reason.* Kant's main concern was to explain how synthetic a priori judgements are possible in science and mathematics.

4. Plato, *Meno and Theatetus.*

5. J. Hintikka, *Knowledge and Belief,* Ithaca, NY: Cornell University Press, 1962.

Hintikka's Definition of Knowledge

Following Hintikka, *knowledge* is defined as *justified true belief* as follows: S *knows* that *p*, if and only if

1. *p* is true
2. *S* believes that *p*
3. *S* has adequate evidence for believing that *p*, and the tacit implication of (iii) in a form in which it came to be known as the K-K thesis, i.e.
4. "*S* knows that *p*" entails that "S *knows* that she *knows* that *p*."

Structure of the Paper

In this paper we would like to show that the Received/ Standard View on Epistemology was first and foremost committed to a propositional representation of knowledge and provided the underlying notion of knowledge for two of the most influential models of scientific knowledge, viz. the Building-block/Foundational model and the Subjective, Relativistic, Socio-historical View of scientific knowledge. A viable naturalization programme in epistemology should take the sub-symbolic grounding of concepts as the basis for both knowledge representation and revision.

The paper is divided into two Parts. In Section I.1, we take a critical look at the Building-block/Foundational model as developed by the Logical Positivists and suggest a new set of reasons for further exposing the weaknesses of their syntactic approach to scientific knowledge and its representation, especially their notion of scientific theory. These new reasons go on to show that the Logical Positivistic approach obscures a number of *cognitively* and *epistemologically* significant features of knowledge representation and revision in science.

Section I.2 reviews the post-positivistic school in philosophy of science, the rise of the subjective, relativistic, socio-historical view of scientific knowledge and its commitment to the propositional representation of scientific knowledge. This view appears to compromise on the objective nature of scientific knowledge and shows a lack of understanding of the *cognitive processes* involved in the interaction between cognitive system and the world particularly at the sub-symbolic or conceptual level, which appears to explain the objectivity of scientific knowledge, although there was a general acknowledgement of cognitive phenomena, both individual and social.

In the context of the demise of the inferential justificatory account of knowledge and the reformulation of the adequacy of evidence condition, Section II of the paper highlights the necessity of the subsymbolic information-theoretic approach and ecological factors forming the basis of a knowledge acquiring system in the naturalization of epistemology programme. In Section II.1, after reviewing the Cognitive Turn in Philosophy of Science and the various approaches that has emerged we elaborate the contributions of Dedre Gentner and George Lakoff on analogy and metaphor and the view known as Experiential Realism.

Section II.2 gives the relevant details of the programme on the naturalization of epistemology relating to the non-linguistic, sub-symbolic, ecological aspects of cognition and information theoretic approach to epistemology based on the works of Fred Dretske, J.J. Gibson, David Marr and Peter Gardenfors. Their contributions highlight the role played by the *sub-conceptual level* in deriving reliable information about the world and providing the basis for the formation of concepts that designate natural properties or natural kinds. However, in this endeavour one notices a clear shift of emphasis away from the propositional representation and revision of scientific knowledge.

Section III shows an alternative approach to revision of scientific knowledge based on the work of Paul Thagard.

The paper concludes that the developments in Cognitive Science provide evidence which go on to show how the information about the way the world is can be represented and revised in terms of frameworks that are non-propositional.

I

The Received/Standard View on Epistemology

At the core of the received/standard view on epistemology were two interrelated notions: (i) the so-called *syntactic notion of theory* and knowledge representation, and (ii) the building-block/foundational model of epistemology.

I.1 THE SYNTACTIC NOTION OF SCIENTIFIC THEORY AND KNOWLEDGE REPRESENTATION

The syntactic notion of scientific theory of the logical positivists was central to their epistemology of science. Carnap[6] developed a notion of scientific theory by combining the Duhem-Poincare view that scientific theories are systems of hypotheses or conventions designed to save the phenomena with the Hilbert formalization programme to the effect that all theories — physical as well as mathematical — can be reconstructed as formal axiomatic systems. Giving the hypotheses or conventions the status of axioms in an intended interpretation of a formal system, the logical positivists construed scientific theories as formal axiomatic calculi in which the theoretical

---

6. R. Carnap, "Foundations of Logic and Mathematics", in *International Encyclopedia of Unified Science*, vol. 1, Chicago: University of Chicago Press, 1939.

terms and statements are given a partial observational interpretation by means of correspondence rules.

The logical positivists thus gave this notion of scientific theory a linguistic turn leading to a two tier *physicalistic language* — observational and theoretical — within a formal axiomatic framework. Although it would be difficult to enumerate the tenets unanimously agreed upon by the philosophers in the positivistic era, the more familiar ones are: (i) canonical formalized structure of scientific theory in terms of a first order language L, a syntactic axiomatic calculus, (ii) distinction between observation and theoretical terms, (iii) correspondence rule and partial interpretation of theoretical terms, (iv) observation and experiment playing a crucial role in the relationship between theory and phenomena, whether it be testing or confirmation of the theory, or in the measurement, prediction or explanation of phenomena, (v) the analytic-synthetic distinction, and (vi) a method of justification of the theory or general laws: induction, confirmation, or falsifiability principle.

In this framework one must distinguish (a) the formal axiomatic system, (b) the interpretation of (a) with respect to a specific domain, and (c) the phenomenon in the domain which constitutes the intended interpretation. To illustrate, the formal axiomatic system in which classical mechanics may be formulated is a linguistic entity that finds interpretation in the phenomena dealt with in classical mechanics, such as the dynamics of the solar system. By identifying scientific theory with the formal axiomatic system, the logical positivists regards the theory as a linguistic entity and distinguishes it from the domain of interpretation, i.e. the phenomenon that stands in a certain relationship with a formal axiomatic system as shown by Tarski. It is this relationship which enables the theory to represent observable phenomena and constitutes a theory of representation.

Alternative to the Syntactic View: Semantic Approach to Scientific Theory

Using the insights of Tarski's semantic techniques of model theory some philosophers of science, such as Suppes, Beth, van Fraassen, Frederick Suppe, and Giere maintain that it is a mistake to think scientific theories as collection sentences or propositions and a theory directly represents the world or the world directly satisfies the theory. Had theory been a linguistic entity, there could not have been different linguistic formulations of the same theory. Moreover, if one takes into account the fact that a typical scientific theory is always appended with *ceteris paribus* clauses, simplifications and idealizations it would be naïve to think that theories are direct representations.

The proponents of the semantic approach made a valiant attempt to liberate the philosophical study of science from its *linguistic dependence* created by the *Received View on Scientific Theories* of the logical positivists. One of the proponents of the alternative view, the *Semantic View of Scientific Theory* view, van Fraassen, says:

> The syntatic picture of a theory identifies it with a body of the theorems, states in one particular language chosen for the expression of that theory. This should be contrasted with the alternative of presenting a theory in the first instance by identifying a class of structures as its models. In this second, semantic approach the language used to express the theory is neither basic nor unique; the same class of structures could well be described in radically different ways, each with its own limitations. The models occupy centre stage.[7]

7. B.C. van Fraassen, *The Scientific Image*, Oxford: Oxford University Press, 1980, p. 44.

According to Frederick Suppe,[8] a convenient starting point for explaining the distinction of the *Semantic View of Scientific Theory* would be the notion of a *physical system.* A scientific theory does not deal with phenomena in all of their complexity, rather it characterizes phenomenal systems by selecting certain idealized parameters from the phenomena in terms of a class of *abstract replicas* of the phenomenal systems, known as *physical systems.* For example, in characterizing falling bodies classical particle mechanics is neither concerned with, say, the colour of the object, nor with motion under normal circumstances. Rather it is concerned with the behaviour of isolated systems of extensionless point-masses which interact in the idealized frictionless environment of vacuum, where the behaviour of these point-masses depend only on their positions and momenta at a given time. Thus, the *physical systems* are highly abstract and idealized replicas of phenomena, being characterizations of how the phenomena *would have* behaved *had* the idealized conditions been met. This is where scientific *hypotheses* and the *counterfactual nature* of scientific laws play a role. The main task of a *scientific theory,* therefore, is to describe, predict and explain the behaviour of a class of physical systems, which correspond to phenomena.

The semantic approach takes a theory to be a class of models. The term "model" is to be understood in the logician's sense: a structure that makes a sentence or a set of sentences (theory) true. Thus, a *model* is given by specifying a structure in which the axioms of a given theory T turn out to be true. This concept of model requires independent determination as to whether any given structure indeed makes the axioms of T true. In this sense a theory is an extra-linguistic abstract structure, which can be described by different linguistic formulations.

8. F. Suppe, op. cit., pp. 221-30.

Once we acknowledge the relevance of this view we begin to see how the *Received View on Scientific Theories* conceal much that is epistemically and cognitively important and fail to reveal the relationship between scientific theories and phenomena. First, the scientific laws, say, Newton's Laws, are not to be equated with a set of sentences or specific linguistic formulations. They are to be interpreted as providing definitions of various physical systems or models that are *non-linguistic,* though *abstract,* entities. Secondly, it is the set or a family of models or the abstract physical systems and the phenomena or real systems that are claimed to be related, based on the *hypotheses* of the theory, and not the set of statements defining the physical system or model and the phenomena or the real system. Thirdly, the scientific hypothesis formulated on the basis of observation and experiment made within the parameters set by the theory attempt to exhibit the *similarity* between the models or the abstract physical systems and the phenomena or real system. Hence, a scientific theory comprises two elements: (a) a family of physical systems or models, and (b) a set of hypotheses claiming similarity between the physical systems or models and the phenomena or real systems and establishing a link between the two. Van Fraassen says:

> To present a theory is to specify a family of structures, its models; and secondly, to specify parts of those models (the empirical substructures) as candidates for the direct representation of observable phenomena.[9]

However, the notion of representational relation between the theory and the world in the semantic approach is a very complex one. Although equations, diagrams, graphs, ball-and-stick structures (especially in Chemistry), abound in scientific

9. B.C. van Fraassen, op. cit., p. 64.

theories, they do not represent the world directly but often describe idealized, unactualized physical systems, *abstract replicas* of the phenomenal systems, known as *physical systems,* whose contents are actual phenomena or real-worldly physical systems. Our acceptance of a theory depends on the extent in which the behaviour of the idealized, unactualized physical system is similar to the actual phenomena.

The Received/Standard View on Epistemology and the Propositional Representation of Knowledge

The Received/Standard view on knowledge clearly endorses the *propositional representation* of knowledge. The reasons for this are quite obvious.

First, according to this view knowledge requires belief. This refers to condition (ii) mentioned above. Secondly, belief is construed as a binary relation involving propositional attitude, i.e. the doxastic attitude of an epistemic agent *S* directed towards a proposition, or the content of the belief, expressed in the form of an embedded *that*-clause, e.g. *S* believes *that the cat is on the mat.* Thirdly, for the belief to be true (i.e. satisfying condition (i) of Hintikka) or for the epistemic agent to have degrees of conviction, confidence or objective probability attributable to a proposition requires that belief or knowledge be propositional. Finally, in order to enable an epistemic agent to attribute truth-value or some objective probability to a proposition, she requires evidence and evaluation leading to justification (i.e. condition (iii) mentioned above). Justification (whether foundational or coherentism) involves argument and arguments consist of proposition. The science practised during the seventeenth through twentieth centuries created new requirements and imposed new constraints on the characterization of evidential support to (scientific) knowledge resulting in different *models of epistemology.*

Modern Science and the Building-block/ Foundational Model of Epistemology

One of the models that is claimed to characterize the underlying epistemology of modern science was *the Building-block/ Foundational Model of Epistemology* developed by some of the eminent members of Logical Positivism.

This resulted in the so-called *building-block/foundational model* in epistemology (mentioned long time ago by Descartes) with a body of *base knowledge* taken as *incorrigible*. The knowledge of any other type, say, of entities not directly observable or generalizations in the form of scientific laws, were to be shown as derivative, i.e. obtained in terms of mental or logical operations from the base knowledge and rendered indubitable or incorrigible by it. The scientific laws can be gleaned from natural phenomena. However, some bedrock of perceptual certainty is necessary for gaining a firm knowledge of the world. The Logical Positivists introduced the doctrine of *cognitive significance* in the form, such as the verification principle of meaning and the principle of falsifiability.

THE RECEIVED/STANDARD VIEW ON KNOWLEDGE AND THE BUILDING-BLOCK/ FOUNDATIONAL MODEL OF SCIENTIFIC KNOWLEDGE

In what measure does the building-block model of scientific knowledge provide insights into our understanding of the defining characteristics of knowledge spelled out by Hintikka? How did these defining characteristics of knowledge spelled out by Hintikka fare in providing insights into our understanding of *scientific* knowledge?

First, according to the building-block model in epistemology, science initially consists of empirical generalizations formulated using *observation* terms. As science advances, *theoretical* terms are introduced by definition and *theoretical laws* are formulated with the help of *theoretical* terms. Hence, science exhibits an upward inductive movement from particular facts to theoretical generalizations, i.e. a body of generalizations from a base in observation and justifying inductive inferences from observation to theory.

The satisfaction of conditions (iii), i.e. adequate evidence condition, and (iv), i.e. the K-K thesis, of knowledge mentioned by Hintikka, is ensured by the building-block model in epistemology, since any *derivative knowledge* in that model is rendered incorrigible provided the *base knowledge* is. The incorrigibility of the base knowledge is explained (although initially by Descartes, in terms of his notion of clear and distinct ideas), by the Logical Positivists, in terms of a *neutral* observation language, such as *sense data statements*. However, the incorrigibility is *preserved* by the logical or mental operations, which enable us to obtain derivative knowledge from the base knowledge. That is, adequacy of evidence or incorrigibility is *transmittable* or *transferable*.

This goal was sought to be achieved, from the very inception of the Logical Positivistic programme, by developing a satisfactory *inductive logic*.

INDUCTION AND THE BUILDING-BLOCK/ FOUNDATIONAL MODEL OF SCIENTIFIC KNOWLEDGE

Carnap[10] assiduously attempted to construct an inductive logic based on *deductive implication*, i.e. observation statements relating to empirical evidence and perceptual data would *deductively imply* law statements and generalizations. This was based on the vision that scientific inferences are both *rational* and *objective* exactly the way deductive logic was taken to be.

10. R. Carnap, *Logical Foundations of Probability*, Chicago: Chicago University Press, 1962.

For his inductive logic, Carnap assumed that (i) scientific language is a version of first order logic and physical theory can be constructed from a base in observation within a formal deductive framework, as Russell and Keynes visualized, (ii) the *atomic predicates* of the observational language are *primitive notions, logically independent* of each other and all such predicates are *to be treated in a similar fashion* as there is no logical reason for differentiating them, and (iii) the relationship between the observation statements and the general law statements is one of *partial implication* explained in terms of *probability function*. Schematically, it amounts to saying that $p(L/e) = r$, i.e. the probability of L, given evidence e, has the value r.

Carnap's version of inductive logic was pursued by many in the 1970s. From the above it can be seen easily that the approaches outlined by Carnap and his followers are not only consistent with Hintikka's conditions (iii) and (iv), but condition (i) as well, as incorrigibility and truth are often used interchangeably.

Reichenbach[11] developed his inductive rule of inference, the so-called *straight rule*, in order to achieve the goal of scientific generalization, viz. determining the limiting relative frequencies in infinite sequences. Although his notion (of limiting relative frequency of attributes in an infinite sequence) has similarities with the structure of probabilities, he construed it as a *norm for practical action* and gave it a *pragmatic* justification. For example, if a fraction, f, of observed marbles are red, then Reichenbach would suggest that we posit, as a matter of pragmatic rule, that the limiting relative frequency of marbles

---

11. H. Reichenbach, *The Theory of Probability*, Berkeley: University of California Press, 1949. See also, W.C. Salmon, *The Foundations of Scientific Inference*, Pittsburgh: University of Pittsburgh Press, 1977.

in the infinite limit should be f, provided the sequence in question does have a limit.

However, all the attempts in this direction failed. This led some, such as Popper, to abandon the inductive logic-based approach to justification for scientific laws in favour of falsificationism, without giving up the main tenets of the epistemology of science as developed by the Logical Positivists.

The Building-block/Foundational Model of Scientific Knowledge obscures the Cognitively and Epistemologically Significant Features of Scientific Knowledge

It seems to me that the reason why inductive logic turned out to be so problematic is not that the relevant technical problems could not be solved or that its use could not be substantiated by actual scientific practices, as argued by Lakatos against Popper, but the fact that *it is not supported by our present understanding of cognition* that accounts for our inductive abilities.

Formal Epistemic Logic vs Cognitively Valid Epistemic Logic

The positivistic analysis of induction ran into serious trouble because of riddles[12] and paradoxes[13] encountered by it. The debates centring around the riddles and paradoxes of induction exhibit that any analysis of *inductive capacity* must eschew the positivistic assumption that all *primitive predicates* are logically on the same par, logically independent of each other and atomic.

---

12. N. Goodman, *Fact, Fiction, and Forecast,* Cambridge, MA: Harvard University Press, 1955. Especially chapter III, section 4, entitled "The New Riddle of Induction".

13. C.G. Hempel, *Aspects of Scientific Explanation and Other Essays in Philosophy of Science,* New York: Free Press, 1965.

The Logical Positivists were concerned with the logical analysis of propositions/sentences in a regimented language whose atomic predicates designate observational properties. They took the observational predicates as primitive, unanalysable notions and treated all observation predicates in the same way, as there were no logical reasons for differentiating between them. A cognitive approach to epistemology that explains how the predicates get established and the cognitive grounds on which distinctions can be made between predicates repudiates the Logical Positivistic assumption mentioned above.

This stricture on the Positivistic approach to primitive predicates can be substantiated by the fact that the common thread, running through all the myriad solutions of the riddles and paradoxes of induction that are still being debated, is the search for the basis in the distinction among the predicates in terms of identification of those predicates which can be used in inductive inferences as opposed to those which cannot. This clearly suggests the unacceptability of the Positivistic assumption mentioned above. Of course, different epistemic and cognitive criteria have been suggested in order to differentiate the predicates that are useful in inductive inference, such as projectibility, entrenchment, natural kind, positional-qualitative solution, simplicity solution, the coherence solution, the incoherence dissolutions, the falsificationist response, the real property approach, the counterfactual approach, and so on. However, these attempted solutions exhibit distrust on the Positivistic assumption about primitive predicates and an increasing sensitivity towards cognitive issues relevant to an analysis of human being's inductive capacities.

Bayesian Inference and the Cognitive Approach to Thinking under Uncertainty

In spite of the failure to come up with any satisfactory account of induction, philosophers of science refused to give up their attempts to explain the nature of scientific inference leading to laws in logical and mathematical terms, i.e. in "reducing scientific rationality to logicality."[14] The logic of probability was construed as the very core of scientific inference as it was taken to provide the methodological basis for the justification of scientific claims.

BAYESIAN INFERENCE

One of the significant attempts in this direction is Bayesian inference and approach to the confirmation of scientific theory. According to this view the initial assumption for confirmation is that our attitude towards theories are measured by the *subjective probabilities* we attach to them, i.e. an evidence E *confirms a theory* T if learning E should make one *increase* her *subjective prabability* for T: if it is a hazardous speculation one's subjective probability would be close to 0; however, if one fully believes a theory her subjective probability would be 1.

Bayesians next introduce the notion of *conditional probability*: the *conditional probability of A given B* (written "Prob(A/B)") is defined as the quotient of Prob(A and B)/ Prob(B), and can be thought of as the probability-of-A-on-the-assumption-that-B-is-true.

Let us assume that E is some possible evidence and T is some theory. Prob(T/E) is then the probability of T, on the assumption that E is true. Bayesians, therefore, argue that when one learns E, one should increase her probability for T

---

14. S. Toulmin, *Human Knowledge*, Princeton, N.J: Princeton University Press, 1972.

to equal this number. Hence, for Bayesians E will confirm T, if and only if Prob(T/E) is greater than Prob(T).

Bayes's Theorem now tells us more about when E will confirm T. The Theorem follows from the definition of conditional probability. According to this definition Prob(T/E) = Prob(T and E)/Prob(E), while Prob(E/T) = Prob(T and E)/Prob(T). Putting these two together, we get

Bayes's Theorem: Prob(T/E) = Prob(T) ? Prob(E/T)/Prob(E).

The Theorem claims that Prob(T/E) is greater than Prob(T), i.e. E confirms T if and only if Prob(E/T) is bigger than Prob(E) or, in other words, E confirms T to the extent that E is likely given T, but unlikely otherwise. For example, if E is in itself very surprising, e.g. light bending in the vicinity of the Sun, but at the same time exactly what one would expect given theory T, viz. the general theory of relativity, then E should make one increase her degree of belief in T a great deal. On the other hand, if E is no more likely given T than it would be on any other theory, then observing E provides no extra support for T. For example, the movement of the tides is no great argument for the general relativity theory, even though it is predicted by it, since it is also predicted by the alternative Newtonian theory of gravitation.

Horwich's Attempt to Solve the New Riddle of Induction by using Bayesian Induction

Paul Horwich[15] is one of the first few who attempted to *express the idea of projectibility in probabilistic terms* and thereby offered a *Bayesian solution* to the *new riddle of induction.*

Horwich calls a hypothesis projectible when the observation of positive and no negative instances tend to

15. P. Horwich, *Probability and Evidence,* Cambridge: Cambridge University Press, 1982.

confirm the conclusion that the unsampled A's are B. He, then, raises the question whether there is a necessary and sufficient projectibility condition on the prior probability of a hypothesis. He argues that if the hypothesis H, viz. *All sampled and unsampled A's are B,* is projectible, then the product of the prior probabilities of the grue-like (or gruesome!) alternative to H must be less.

This shows that projectibility is associated with high prior probability and concludes that in most cases the probability is approximately that not all sampled and not all unsampled A's are B, and so if P(H) is greater than the probability that all sampled and not unsampled A's are B or the probability that not all sampled and all unsampled A's are B, then H is projectible.

Horwich observes that this works with hypotheses that have high prior probabilities, but not with hypotheses that have low prior probabilities, such as the grue hypothesis, the observation of whose positive instances we feel provides no reason to believe that they will be satisfied in the future. On a Bayesian approach if two hypotheses, which explain the data equally, will nevertheless have different posterior probabilities, then they must have had different priors too. The grue hypotheses have lower prior probabilities than the green hypotheses. This shows that although the consequence condition is false it is a central element of scientific methodology. Thus, according to Horwich, the new riddle reduces to the problem of describing the basis of our prior probability assignments.

If projectibility is based on the notion of relatively high prior probability then it is necessary to explain how degrees of belief are measured and probability claims are interpreted in an *objective* manner. Moreover, Horwich's account of projectibility has counter-intuitive results. For example, the hypothesis *all A's are B* can be projectible even though the prior probability that all sampled A's are B is *much* lower than the prior probability that all sampled A's are B's and not all unsampled A's are B's and *much* lower than the prior probability that not all sampled A's are B's and all unsampled A's are B's.

Cognition and Thinking under Uncertainty

There are several reasons for dissatisfaction with the Bayesian approach to the theory confirmation: first, the initial measures supplied by the individuals, viz. the notion of subjective probability, smacks of relativism. Secondly, the assumption that the agent knows all the options logically and physically open to her, and all the logical and physical states that might be relevant to her for taking a decision, as well as the assumption that the agent has a coherent preference structure and is able to maximize based on the calculation of the expected utility for each option, are, to say the least, too idealistic. Thirdly, given the fact that there are many different logically possible ways of *conditionalizing* on the evidence, there is no a priori way of making a choice, which may be regarded as uniquely rational, out of all these alternatives. Fourth, current empirical research on human probability judgement show that the canons of classical theory regarding probability are too idealistic, do not take cognizance of the cognitive infirmities of man, and fail to provide realistic, cognitively sound and ecologically valid set of rules reflecting such features of real cognitive systems as boundedness of rationality,[16] biases in probabilistic judgements revealing heuristics of thinking under uncertainty and the failure of human reasoners in arriving at

16. H.A. Simon, *Models of Man*, New York: Wiley, 1957. See also, H.A. Simon, *Models of Bounded Rationality*, 2 vols., Cambridge MA: MIT Press, 1983.

the normatively correct conclusions[17] laid down by Probability theory.

The two well-known problems discussed by Kahneman and Tversky, viz. the *Linda* and the *Taxi*,[18] show that as opposed to the expert trained statisticians, the norms/ standard rules of probability theory are often not adhered to by the lay intuitive novice statistician. The Linda problem shows that the subjects often ignore the rule that the probability that a pair of events, A and B, both occurring cannot be greater than the probability of A occurring with or without B conjunction. The taxi problem again demonstrates the violation of a norm in Bayesian inference that the immediate evidence bearing on some particular case must be weighed against the background rates at which various cases of that sort arise.

Idealistic Logic vs Realistic, Cognitively Sound, Ecologically Valid Logic

From an analysis of these problems it would be wrong to conclude that the lay statistician is *irrational* and lacks in *reasoning ability*, and hence, following Quine, we must be content with only the psychological description. Quine argues that epistemology should be reduced to psychology as psychology alone is competent to tell us what cognitive procedures are involved in formation and justification of our beliefs.

My contention is that there is no reason to give up entirely

---

17. R. Nisbett and L. Ross, *Human Inference : Strategies and Shortcomings of Social Judgement*, Englewood Cliffs, N.J: Prentice-Hall 1980. See also D. Kahneman, P. Slovic, and A. Tversky (eds.), *Judgement Under Uncertainty: Heuristics and Biases,* Cambridge: Cambridge University Press, 1982.

18. R. Nisbett and L. Ross, 1980, op. cit.; Kahneman and Tversky, p. 156.

the normative approach in favour of a purely descriptive one. Rather, we should look for an epistemology, which is based on realistic, cognitively sound and ecologically valid norms.[19] The contrast between the naïve and the expert scientist shows that there are many "heuristics," schemata, models of reasoning. The application of anyone of these in a given situation depends on the knowledge of the appropriateness of these in that specific situation. This is the hallmark of the expert. Lack of such knowledge and "biases" that characterize the naïve scientist explains her proneness to grossly mistaken judgements and errors.

General Remarks Against the Formal Linguistic and Propositional Approach

A close look at the vast literature surrounding the debates mentioned above exposes certain inherent limitations of Logical Positivistic brand of philosophy of science. The most significant limitations are:

(A) There are two most important cognitive issues that are revealed by the *Semantic View of Scientific Theory*. The new notion of scientific models and theories —

(i) brings out the nature of the *representation* used in science. Contrary to the claims made by the proponents of the linguistic and propositional representation the new notion of scientific models and theories discloses that there can be a non-propositional mode of representation, and

(ii) provides a deeper understanding of the nature of *reliable cognitive processes, conceptual devices* and *human*

19. A. Gupta, and N. Sutaria, "Executable Justificational Rationality as Naturalized Epistemology", in, D. Andler et al. (eds.), *Facets of Rationality*, New Delhi: Sage Publications, 1995, pp. 312-44.

judgements that aim at discovering how the world really is and the characteristic regularities of the real world.

(B) Problems with Received View on Knowledge

(i) *The Gettier Problem:* The Received/Standard view on knowledge and the building block/foundational model of scientific knowledge met with a serious challenge to the definition of *knowledge as justified true belief* satisfying conditions (i) to (iii) laid down by Hintikka, showing that the definition is open to a series of paradoxes formulated in 1963 by Edmund Gettier. Gettier paradoxes essentially demonstrate that

(ii)* It is possible for *S* to have adequate evidence for believing that *p* when '*p*' is false, and

(iii)** For any proposition *p*, if *S* has adequate evidence for believing that *p*, '*p*' entails '*q*', and *S* accepts *q* as a result of this deduction, then *S* has adequate evidence for believing that *q*.

Contrary to the Received/Standard view on knowledge, Gettier type paradoxes show that justified true belief may not amount to knowledge, because a belief

(a) can be true and yet not be knowledge as in the case of conjecture, lucky guess or chance, where no amount of internal inspection will help to distinguish the true from the false belief,

(b) may be entertained by an epistemic agent S which makes no reference to truth, or, in other words, truth may not be believed,

(c) can be true and justified, but acquired through a *belief-forming process* that *accidentally* produces true belief, rather than a process that *invariably* and *generally* produces true belief and *guarantees* its truth and avoidance of error, and

(d) cannot be incorrigible satisfying the *adequacy evidence* requirement (i.e. condition (iii) mentioned above), based on correspondence notion of truth as was claimed in the case of sense-data statements, since it would lead to acceptance of private language.

The Gettier problem has spawned an enormous amount of literature[20] and resulted in weakening of the *adequacy evidence* requirement so that the evidence need not be sufficient to guarantee the truth of *p*, but would make it highly probable or likely. Ayer,[21] Chisholm[22] and Malcolm[23] advocated such measures in order to salvage the *received/standard* epistemological view. However, all these attempts to come up with the notion of justification or testability that is compatible with the falsity of *p* were susceptible to some new formulation of the Gettier paradox and the debate remained inconclusive.

(ii) There are several shortcomings that can be enumerated having to do with the general Logical Positivistic approach: (a) on its insistence on the solution of all problems within the framework of a *regimented language* and *propositional formulation*. Even an arch critic of Logical Positivism, Quine, maintains that it is the sentence as a whole (especially *observation sentence*) that is directly correlated to sensory stimulations, and not

20. R.K. Shope, *The Analysis of Knowing: A Decade of Research*, Princeton, N.J: Princeton University Press, 1983.

21. A.J. Ayer, *The Problem of Knowledge*, Baltimore, Md.: Penguin, 1956, chapters 1 and 2.

22. R. Chisholm, *Perceiving*, Ithaca, N.Y.: Cornell University Press, 1957.

23. N. Malcolm, *Knowledge and Certainty*, Englewood Cliffs, N.J: Prentice-Hall, 1963.

single words of the sentence since the reactions of assent and dissent caused by sensory stimulations are directed towards sentences and not towards words. Assent or dissent concerns sentences and not words. According to Quine the contact between language and sensory stimulations is a *holophrastic* one; (b) reducing *scientific rationality* to an idealistic, humanly unattainable, cognitively unfeasible notion of *logicality*; (c) the poverty of a cognitively viable and historically accurate analysis of how *scientific observations* and *experiments* relate to processes of *concept formation* and *learning*, especially with reference to their analysis of *observation terms* and *observation language;* and (d) failure to provide an adequate analysis of *representing scientific knowledge* and *its revision*.

I.2 SUBJECTIVE, RELATIVISTIC SOCIO-HISTORICAL VIEW OF SCIENCE

The subjective, relativistic socio-historical view, advocated very forcefully by Toulmin, Hanson, Kuhn and Feyerabend, also subscribe to the propositional/statement approach to science and construes scientific theories as linguistic entities. According to this view science and scientific theories are collection of propositions/statements whose meanings are determined by, and specific to, a scientific world-view or *Weltanschauung/ Lebenswelt* peculiar to the theory.

The role of the scientific world-view or *Weltanschauung/ Lebenswelt* on the epistemology of science is significant as it determines the class of legitimate problems, delimits the standards for their acceptable solution, and specifies the epistemic grounds involved in the historical and sociological factors responsible for the discovery, development and acceptance or rejection of scientific theories.

Belief/Theory Revision: Growth of Scientific Knowledge

The main emphasis within the subjective, relativistic socio-historical view on Philosophy of Science was to study the *internal* and *external* historical factors responsible for theory change and the epistemological and ontological theories that provide justification for such changes. Some of the most influential work in this area will include: Popper on Conjecture and Refutation, Kuhn's work on Normal and Revolutionary Science, Lakatos on the Methodology of Scientific Research Programme (MSRP), Toulmin's thesis on evolutionary analogy for conceptual evolution.

However, in contrast with a *static model* of human knowledge the proponents of the the subjective, relativistic socio-historical view of science provided a dynamic account of scientific knowledge.

Modern Science and the Subjective, Relativistic Socio-historical View

Although the proponents of the Subjective and Relativistic View differ on many specific issues, their characterization of the scientific enterprise and epistemology have several features in common :

(a) *Scientific observation is theory-laden:* Granting that the objects that are observed and the properties possessed by them are independent of the observer, but, according to this view, what kind of objects they are observed to be and the properties they are observed to have are ***relative*** to and determined by the observer's theory, belief, knowledge, i.e. observer's *Weltanschauung*. Hence, the adherents of two different theories will view the same phenomenon in two different ways.

For Kuhn disciplinary matrices or *paradigms* are constitutive of nature. Hanson maintains that Tycho and Kepler would see different things while watching the same dawn.

This led Scheffler to conclude that *Weltanschauung* analysis leads to *subjective view* of scientific knowledge, which is contrary to the belief that it is objective.[24]

(b) *The meanings of the non-logical descriptive terms* — both *observational and theoretical* — are *theory-dependent*: meaning of a descriptive term depends upon the theoretical context in which it occurs. A term does not mean something in isolation. Its meaning is derived by being part of a theoretical system. The meaning of no term, say *mass*, common in two different theories, for example in Classical mechanics and Special Theory of Relativity, will have the same meaning as different and incompatible equations about mass hold in two theories.

Hence, the condition of *invariance of meaning* across the theories does not hold and different theories are non-comparable or *incommensurable*.

(c) *Underdetermination of theory by facts*: the same set of facts may be explained by more than one theory. What counts as a fact is determined by the *Weltanschauung* associated with a theory.

(d) *Rationality and Theory Change or Growth of Scientific Knowledge*: since 1960s the static Positivistic model of scientific knowledge was replaced by the *dynamic model*. This led to a historical approach and an analysis of the actual scientific practices in order to develop an understanding of the principles that govern rational transitions or justifiable change of belief in the growth of scientific knowledge.

---

24. I. Scheffler, *Science and Subjectivity*, Indianapolis, Ind.: Bobbs Merrill, 1967.

However, the philosophers subscribing to the *Weltanschauung* analysis based on their detailed historical studies came to uphold the view that there is no uniquely correct methodology of science as there exists plurality of methodologies which are employed for all sorts of reasons. Moreover, there is no **rational** criteria for choice between rival theories and the adoption of one methodology over another is not governed necessarily by any rational factor.

(e) *Relativism*: The incommensurability thesis stated in (b) implies *relativism* : the acceptance of one theory over another is relative to prior commitment to a paradigm or *Weltanschauung*, rather than to a neutral, objective criteria of evaluation. The *Weltanschauung* analysis not only reduced scientific knowledge to a subjective enterprise, but made it ***relative*** to a sociocultural group.

The Received/Standard View on Knowledge and the Subjective, Relativistic Socio-historical View of Science

The socio-historical approach to the evolving nature of scientific knowledge takes the canons for evaluating what to admit as scientific knowledge not only as relative to a given *Weltanschauung*, but also as variable from one historical period to another. Although the canons of rationality and justification change and are relative, the conditions characterizing the nature of knowledge, as spelled out in the Received/Standard View on Knowledge, i.e. conditions (i) - (iv) mentioned above, construing knowledge as justified true belief, remain the same.

In spite of its role in exposing the deficiencies of a static, ahistorical and extreme empirical nature of scientific knowledge of the Logical Positivists and replacing it by a more historically sensitive account of scientific knowledge based on an analysis of actual scientific practices, the *Weltanschauung* analysis led to an extreme epistemic relativism. The epistemic view turned out to be so permissible as to accept anything as knowledge as long as any science permits it to enter into its domain and to take any change in the canons leading to corresponding change in what counts as knowledge. This goes against the very basic notion of science as an enterprise providing objective knowledge.

II

The Cognitive Turn in Science Studies

The so-called *cognitive turn* in science studies has resulted due to the following important realizations:

1. that both the words *science* (which comes from the Latin word *sciencea* meaning *knowledge*) and the word *cognition* (co = together, gnocere = knowledge) have *knowledge* in common. Science is taken to be a major paradigm of a knowledge-producing enterprise and Cognitive Science studies the underlying mechanisms responsible for the production, acquisition and deployment of knowledge, including scientific knowledge.
2. that the exclusively a priori, speculative and normative approach to traditional epistemology is to be replaced by an *empirical* and *naturalistic* approach treating cognitive activities and phenomena as natural phenomena. The traditional epistemology has undergone radical changes as it became apparent that an important resource for a naturalistic account of knowledge is Cognitive Science.
3. these mechanisms presuppose that the human mind or artificial intelligent devices can function based only on *representational structures* and *processes* that operate on them to produce new structures. These structures

may include linguistic expressions, concept trees, schemas, cognitive maps, mental models, diagrams, visual images, etc.

In the traditional epistemology, Kant, for example, provided such an account of the underlying mechanism for the possibility of scientific knowledge by adopting a speculative and a priori approach. Kant asked the question as to what sort of mechanism would make mathematical and natural scientific knowledge possible and gave a description of such a mechanism.

In the 1950s, Kuhn and Hanson used psychological ideas, particularly from the Gestalt school, in order to explain certain phenomena in scientific observation. However, due to the progress made in Cognitive Science since 1950s there is a better understanding of the cognitive mechanisms involved in many scientific activities and practices.

A survey of the relevant literature shows that the Cognitive Science-based approach *complements* the traditional historical approach (both internal and external), the sociological approach, and the philosophical concern for the epistemology of science, by providing an enriched understanding of how scientists generate and evaluate scientific ideas. For example, the volume (no. XV), entitled *Cognitive Models of Science*, published in 1992 by R.N. Giere[25] in the series "Minnesota Studies in the Philosophy of Science" presented the work of psychologists, historians, researchers in Artificial Intelligence and philosophers who had employed ideas and methodologies from Cognitive Science in order to study the underlying cognitive mechanisms for explaining and understanding scientific practices.

---

25. R.N. Giere (ed.), *Cognitive Models of Science*, Minneapolis: Univ. of Minnesota Press, 1992.

Another notable effort to understand what features of the human mind and of the human culture and cognitive development permit, facilitate and make science and the conduct of science possible has been undertaken by the University of Sheffield's Hang Seng Centre for Cognitive Studies and in their various publications, most significant among them is the volume entitled *The Cognitive Basis of Science.*[26]

The earlier work by Kuhn, Hanson, Campbell, Quine, Bruner, Goodnow and Austin, and Simon strongly suggest that the gap between Philosophy of Science and Cognitive studies needs to be closed. Since their pioneering work the emerging multi-disciplinary nature of Cognitive Science has been providing valuable resources for the philosophical study of science just as historical data did. Some of the important examples of Cognitive-historical approach to Philosophy of Science are: Giere (1988, 1991),[27] Fodor (1983, 1984),[28] Churchland (1979),[29] Nersessian (1984),[30] Thagard (1988, 1991),[31]

---

26. P. Carruthers, S. Stich and M. Siegal (eds.), *The Cognitive Basis of Science*, Cambridge: Cambridge University Press, 2002.

27. R. Giere, *Explaining Science: A Cognitive Approach*, Chicago: University of Chicago Press, 1988. See also, R. Giere, *Understanding Scientific Reasoning*, 1991, 3rd edn.

28. J. Fodor, *The Modularity of Mind*, Cambridge, Mass.: MIT Press, 1983. See also J. Fodor, "Observation Reconsidered", *Philosophy of Science*, 51 (1984), pp. 23-43.

29. P.M. Churchland, *Scientific Realism and the Plasticity of Mind*, Cambridge: Cambridge University Press, 1979.

30. N.J. Nersessian, *Faraday to Einstein: Constructing Meaning in Scientific Theories*, Dordrecht: Nijhoff, 1984. See also N.J. Nersessian (ed.), *The Process of Science: Contemporary Philosophical Approach to Understanding Scientific Practices*; N.J. Nersessian (ed.), *Selected Works of H.A. Lorentz.*

31. P. Thagard, *Computational Philosophy of Science*, 1988. See also P. Thagard, *Conceptual Revolutions*, 1991.

Goldman (1986),[32] Kahneman et al. (!982),[33] Lakoff (1987, 1980),[34] Gardenfors (2000),[35] Gentner (1983, 1983),[36] Carey (1985),[37] and Keil (1979, 1989).[38]

Moreover, there are major studies on scientists and scientific theories from the cognitive point of view. A few examples of such studies may be mentioned here. Darden[39] has identified the cognitive strategies that contributed towards the development of Mandelian genetics. Giere has used psychological and sociological ideas to improve our understanding of the recent developments in geology and

---

32. A. Goldman, *Epistemology and Cognition*, Cambridge, Mass.: Harvard University Press, 1986.

33. D. Kahneman, P. Slovic and A. Tversky (eds.), *Judgement under Uncertainty: Heuristics and Biases*, Cambridge: Cambridge University Press, 1982.

34. G. Lakoff, *Women, Fire, and Dangerous Things*, Cambridge: University of Chicago Press, 1987. See also G. Lakoff, with M. Johnson, *Metaphors We Live By*, Chicago: University of Chicago Press, 1980.

35. P. Gardenfors, *Conceptual Spaces: The Geometry of Thought*, Cambridge, Mass.: MIT Press, 2000.

36. D. Gentner, "Structure-mapping: A theoretical framework for analogy", *Cognitive Science*, 7 (1983), pp. 155-70. See also D. Gentner and D.R. Gentner, "Flowing Waters or Teeming Crowds: Mental Models of Electricity", in D. Gentner and A.L. Stevens (eds.), *Mental Models*, Hillsdale, N.J.: Erlbaum, 1983, pp. 99-129.

37. S. Carey, *Conceptual Change in Childhood*, Cambridge, Mass.: MIT Press: 1985.

38. F.C. Keil, *Semantics and Conceptual Development*, Cambridge, Mass.: Harvard University Press, 1979. See also F.C. Keil, *Concepts, Kinds, and Cognitive Development*, Cambridge, Mass.: MIT Press, 1989.

39. L. Darden, *Theory Change in Science: Strategies from Mendelian Genetics*, Oxford: Oxford University Press, 1991.

physics. Based on the case study on plate tectonics, Solomon shows how certain cognitive heuristics play a crucial role in scientific decision making. Nersessian has drawn on ideas from cognitive psychology to help understand the developments in electromagnetic theories made possible by the contributions of Faraday, Maxwell and Einstein. Churchland has discussed the nature of theories and explanations from the perspective of computational neuro-science. Thagard has used computational and cognitive theories to help understand the structure and growth of scientific knowledge.

APPROACHES TO STUDY OF SCIENCE FROM COGNITIVE SCIENCE POINT OF VIEW

Cognitive Scientists have developed several approaches to issues in Philosophy of Science based on certain assumptions:

(1) One of the earliest approaches in Cognitive Science based on Newell-Simon's Physical Symbol System Hypothesis regards rational thought as consisting of computation or manipulation of abstract symbols, which derive their meaning by representing things in the database or the external world. Intelligent thinking essentially amounts to mechanical means-end procedure performed on symbolic representations. Thinking thus is purely logical, abstract and disembodied capable of being instantiated in appropriate physical substrates.

Fodor[40] and Pylyshyn have argued that some sort of mental code in the form of symbol-strings (mentalese) capture what mental phenomena, such as thoughts,

40. J.A. Fodor, *The Language of Thought*, Thomas Y. Crowell, New York, 1975. See also J.A. Fodor, and Z.W. Pylyshyn, "Connection and Cognitive Architecture: A Critical Appraisal", *Cognition*, 28 (1988), pp. 3-71.

beliefs, are. Their view is known as Language of Thought (LoT).

Simon studied scientific discoveries by developing computer model, e.g. BACON programme, which was designed to simulate actual discoveries from history of science, but capable of making actual discoveries as well in laboratory experiments with subjects in discovery situations. Simon made extensive use of production system programming languages for cognitive simulation.[41]

(2) Cognitive Neuroscience (CN) has provided an alternative research paradigm for Cognitive Science in contrast with the symbol manipulation approach. CN uses several non-evasive techniques, such as positron emission tomography (PET), nuclear magnetic resonance imaging (MRI), functional MRI (fMRI), electroencephalography (EEG) and event-related potential (ERP), which can provide electrophysiological indices for spatial and temporal distribution of brain activities, known as *event related potentials,* viz. measurement of blood flow and the local brain metabolism. These regional brain activities reflect and are associated with such cognitive tasks as memory, attention, solving problems, and recognizing objects. The earliest impetus for CN came from brain pathologies or systemic breakdowns, such as Alzheimer's disease.

---

41. P. Langley, H.A. Simon, G.L. Bradshaw, and J.M. Zytkow, *Scientific Discovery: Computational Explorations of the Creative Processes*, Cambridge, Mass.: MIT Press, 1987. See also D. Kulkarni, and H.A. Simon, "The Process of Scientific Discovery: The Strategy of Experimentation", *Cognitive Science*, 12 (1988), pp. 139-76.

(3) Another very important research paradigm is the Connectionist approach.[42] In contrast with the approach described in (1), which consists of manipulation of abstract symbols using in a serial manner explicit logical rules arranged in a hierarchy, the Connectionist approach invokes parallel processing of sub-symbols. Connectionism attempts to explain cognitive abilities using artificial neural networks. These networks are simplified models of the brain, assumed to be analogs of neurons and composed of (i) large number of units differentiated in three layers, viz. input, hidden and output units and (ii) the weights that measure the strength of connections between the units. The central goal of the Connectionist research paradigm is to find the right set of weights in terms of learning algorithms for performing a given cognitive task. Backpropagation is the most favoured training method, which consists of adjusting the algorithm and the training set. Using this method a net can be trained, for example, to perform the task of differentiating handwritings or faces. While training a net to distinguish a male face from a female face the weights are adjusted in such a way that the value of "male" output unit is increased and that of "female" output is decreased. Thus the weights are numerical values: positive values are excitatory and negative values are inhibitory. After this process of obtaining positive values by adjustment is repeated many times, the net may learn to produce the desired output. The two major advantages of

42. D.E. Rumelhart, "The Architecture of Mind: A Connectionist Approach", in P. Thagard (ed.), *Mind Readings*, Cambridge, Mass.: MIT Press, 1998. See also W. Bechtel and A. Abrahamsen *Connectionism and the Mind: Parallel Processing, Dynamics, and Evolution in Networks*, 2nd edn., Oxford: Blackwell, 2001.

Connectionist modelling are its emphasis on (a) distributed and (b) sub-symbolic modes of representations. By distributed representation one means that instead of using a single unit, say, a neuron firing when one thinks of grandmother, Connectionism associates a complex pattern of activity distributed across a relatively large part of cortex, mathematically a n-dimensional vector. Using numbers it also allows one to define similarity. Moreover, Connectionism espouses sub-symbolic representation by rejecting the view that complex representations are composed out of simple atoms.

Paul Churchland has championed a connectionist neuro-computational theory as a source of answers to traditional problems in philosophy of science,[43] e.g. nature of scientific theory, the problem of theory change in science. According to the Logical Positivists, the Received view on epistemology and LoT, scientific theory is a body of sentences and the knowledge of the theory amounts to storing of these sentences. From Churchland's neuro-computational perspective language is displaced both as a medium of thought as well as scientific theorizing as he believes that the fundamental mode of human cognition is *non-linguistic.* For Churchland to learn a theory or a *paradigm* is for the brain to have acquired a configuration of synaptic weights that partitions a vector space according to similarity relations.

43. P. Churchland, "On the Nature of Theories: A Neuro-computational Perspective", in W. Savage (ed.), *Scientific Theories*, Minneapolis: University of Minnesota Press, 1989. See also P. Churchland, *A Neurocomputational Perspective: The Nature of Mind and the Structure of Science*, Cambridge, Mass.: MIT Press, 1989.

(4) Another emerging approach in Cognitive Science has been represented by a cluster of groups, who emphasize the ecological structure, gestalt properties rather than atomic, interactive, bodily experience as the basis for cognitive systems, non-linguistic, sub-symbolic aspect of cognition and information theoretic approach. This is a heterogeneous group. However, following Lakoff it can be called *Experiential Realism.* An important contribution of this approach, represented by Gentner and Lakoff, would be on analogy, metaphor and categorization, including science (Section I). The main advocates of the sub-symbolic and ecological approach would include Dretske, Harnad, Barwise and Perry, J.R. Anderson, J.J. Gibson, David Marr, Lakoff, Gentner, Langacker, Gardenfors (Section II). In Section III we discuss a concept-based approach to revision of scientific knowledge.

II.1 Experiential Realism and Philosophy of Science

Experiential Realism denounces the disembodied transcendental nature of knowledge and its sentential representation advocated by the Received View. It claims that knowledge goes beyond mere internal coherence and grows out of motor movement, perception, experience, interaction with a stable external world and social environment.

I shall discuss the implications of Experiential Realism to Philosophy of Science in terms of the works of (a) Gentner and Lakoff on analogy and metaphor and (b) Dretske, Gibson, Marr, Harnad and Gardenfors on the non-linguistic, sub-symbolic aspect of cognition and information and concept formation.

ANALOGY AND SCIENTIFIC THINKING

Analogy requires discernment of *similarity* or *agreement*

between things. An analogy promotes better understanding of a phenomenon in terms of another phenomenon, which is familiar, better understood, and more easily or readily grasped. Suppose there is a *similarity* between A and B, and the principles governing *A* are wellknown, then it would be reasonable to expect that *same principles* will govern B, a new phenomenon, which is not understood or a phenomenon is being investigated for the first time.

The ancients attempted to provide explanations for the origin and functioning of nature (e.g. the rising and setting of celestial bodies, lightning and thunder, etc.), the phenomena they did not understand too well, based on analogies between the processes of non-human nature and the functioning of the human body and human society. The analogy invoked (i) self-movement characteristic of living organism to explain such natural phenomena as motion of material objects as the entire nature was thought to be alive, reproductive and self-moving, and (ii) human experiences, such as sympathy and antipathy, attraction and repulsion, love and strife, which were construed as forces that could produce effects on natural phenomena.

In modern science analogical reasoning played an important role in scientific investigations. Huygens, the originator of the wave theory of light, suggested an analogy between the waves of water and sound and those of light. In Maxwell's days hydrodynamic phenomena were better understood than electrical ones. Hence, Maxwell thought it enlightening to compare the latter with the former. In the early decades of the twentieth century an analogy was drawn between the structure of the atom and the solar system in which the nucleus of the atom was likened to the sun and the electrons played a similar role as the planets revolving about the sun. In the Kinetic Theory of Gases an analogy is drawn between gas in a container and the billiard balls, where the molecules in the gas are likened to perfectly elastic billiard balls striking the sides of the container as well as each other. William Harvey in his explanation of the circulation of blood drew an analogy between the heart and a pump.

One of the important questions in the analysis of analogical reasoning is: what sort of similarity will form the basis for an analogical reasoning? In scientific analogy at least the following form the basis:

(i) Similarity in geometrical configuration in the physical structure of the two phenomena under consideration. An atom is considered to be analogous to a planetary system, because in both a relatively light object revolves in an orbit about a relatively heavy one situated in the middle. Huygens saw the same geometrical shape and physical structure of a wave in the light and sound phenomena, on the one hand, and the ripple of water, on the other.

(ii) Similarity by virtue of satisfying the same physical principles. For example, both gas molecules and perfectly elastic balls obey the same conservation laws of classical mechanics.

(iii) Similarity in terms of function or role in two physical systems, such as heart and pump, the gill and the lung in a respiratory organ.

DEDRE GENTNER ON ANALOGY

It is believed that many of the problems in analogy and metaphor interpretation can be dealt with using a mapping known as Gentner's *Structure Mapping* Model,[44] i.e. a familiar situation — the base or source domain — is used as a model for making inferences about an unfamiliar situation — the

44. D. Gentner, 1983, op. cit.

target domain. In the structure mapping the mapping process involves a *structural alignment* between the two domains and the *projection of inferences* from one to the other.

Gentner introduces two important constraints on structural alignment to enable us to make inferences from the familiar (source domain) to the unfamiliar (the target domain). The alignment must (a) be *structurally consistent,* i.e. the mapped elements in the two domains must be in one-to-one correspondence including the arguments of the corresponding predicates, known as *parallel connectivity,* and (b) satisfy *systematicity principle,* i.e. those predicates would be preferred that strongly contribute to the connectivity of the overall structure than those which add little. For example, subject-predicate attribution will have less preference to relational predications of higher arity (i.e. greater number of arguments a predicate or relational symbol can take), which connect many sub-structures and exhibit much grater systematicity. Causal relations would satisfy more the systematicity principle as it contributes more to a richly linked system than an equal number of independent matches.

The Bohr-Rutherford analogy between the conventional solar system and the structure of atom establishes a correspondence not only between sun and nucleus and planets and electrons but also a force of attraction between sun and planets that enables us to make inferences of a similar force which will cause the electrons to stay in orbit and not allow them to fly on a tangent.

Gentner's approach, however, emphasizes, as the very concept "structure mapping" suggests, the structural or syntactic properties of a domain over issues of semantics or pragmatics. In her recent work Gentner is trying to remedy this by moving away from a wholly syntactic direction to more semantic-oriented approach. A good example of such a mix can be found in the work of Lakoff.

LAKOFF ON METAPHOR AND CATEGORIZATION

Lakoff[45] maintains that our knowledge is organized by means of structures called *Idealized Cognitive Models* (ICMs), such as frames, schema, or *metaphoric mappings*. ICMs also explain how people categorize.

Traditionally metaphor was considered as a matter of language or words and not thought. Lakoff's approach to metaphor is not linguistic, i.e. as a rhetorical device that decorates speech. Rather, he explores the cognitive role of metaphor and traces its origins in the conceptual architecture. Metaphors pose a problem for both linguists and philosophers of language. For the transformational linguists metaphors are not part of syntactic deep structure in formal models of language and involves the historical process of meaning change. Therefore, it must belong to the domain of performance rather than competence. Moreover, metaphors force us to consider what it is for a sentence to mean something other than what it says, i.e. its non-literal meaning. The issues relating to non-literal meaning concerns the philosopher of language.

As a generative semanticist Lakoff's interest lies in the cognitive aspect of metaphors: What are the generalizations that govern metaphorical expressions? How do they facilitate the organization of our thoughts and structure our language? Lakoff thinks that the locus of metaphor is not in language but in the way we conceptualize one mental domain in terms of another, i.e. metaphors constitute general mapping (in the mathematical sense) across conceptual domains — from a source domain (i.e. the vehicle) onto a target domain (i.e. the

45. G. Lakoff, *Women, Fire, and Dangerous Things*, 1987.

tenor), thereby causing the target to become grounded in spatio-physical experience via the source.

Lakoff calls such mapping *mnemonics*. To illustrate such mapping Lakoff takes the example of a love relationship that has hit a dead-end, the *love-as-journey* metaphor. He maintains that the mapping is tightly structured. There are ontological correspondence: the entities in the domain of love (e.g. the lovers, the relationship between lovers, their common goals, their difficulties in the love relationship, etc.) match systematically with the entities in the domain of journey (e.g. the travellers, the vehicle, the common destination, impediments in travel, etc.). Thus, the mapping indicates precisely how love is being conceptualized as a journey and enables us to make the following generalizations:

- *Polysemy generalization*: A generalization over-related senses of linguistic expressions, e.g. dead-end street, crossroads, stuck, not going anywhere, etc.
- *Inferential generalization*: A generalization over inferences across different conceptual domains.

In his book, *Women, Fire, and Dangerous Things*, Lakoff also tries to show how ICMs can explain how people categorize. The very title of the book dramatizes and highlights the chain of inference — from conjunction to categorization to commonality as the norm for putting things together based on what they have in common. Instead, in the book Lakoff adopts a cognitive approach to the study of categorization rather than the Fregean objective model of categorization in terms of necessary and sufficient conditions. His ICM approach to category leads to *radial notion of category* where some members of a category are more representative than others. The members of a category thus forms a radial structure, the most representative or prototypical members located in the centre with less representative outsiders clustered around the hub. This also enables us to take care of the borderline cases and make fuzzy inferences.

II.2 The Non-linguistic, Sub-symbolic, Ecological Aspects of Cognition and Information Theoretic Approach to Epistemology

The recent developments indicate that a viable epistemology of science not only needs to take into account the actual scientific practices as revealed by socio-historical studies, but also establish linkages with Cognitive Science and its programme of *naturalization of epistemology*. This may provide a way out of the impasse due to subjectivism and relativism to which the *Weltanschauung* analysis reduced the contemporary philosophy of science and its epistemology. Information Theoretic Approach suggests a viable way to the naturalization programme. We will review the works of Dretske, Gibson, Marr, Harnad, Gardenfors and Thagard.

Naturalization of Epistemology and the Demise of the Inferential Justificatory Account of Knowledge: Adequacy of Evidence Condition and Cognitive Science

Why the information theoretic approach is relevant in epistemology? We have already seen that the Gettier paradoxes posed a serious challenge to the Received/Standard View on Knowledge, especially to its condition (iii), viz. the adequacy condition, which requires that adequate evidence for believing *p* must guarantee the truth of *p*. The Building-block/Foundational Model of Scientific Knowledge tried to fulfil this condition by assuming that (a) the base propositions of our edifice of knowledge refer to sensations and are *incorrigible* and (b) all other propositions about knowledge of the physical world are obtained *inferentially* from (a). The private language argument and paradoxes of induction dealt a deathblow to (a) and (b) respectively and forced us to eventually give up the Building-block/Foundational Model and the inferential justificatory account of knowledge.

At this point one possible strategy for blocking the Gettier paradoxes seemed to compromise the rigid demands implicit in (iii) and accepting the condition (iii), which amounted to saying that it is possible for *S* to have adequate evidence for believing that *p* when "*p*" is false. Ayer and Chishlom adopt such a strategy. However, we observed that this compromise amounts to sacrificing the essence of the Received/Standard View on Knowledge.

The historical account of the development of science show that despite sceptical warnings the success of scientific investigations enabled us to gather much useful knowledge about the universe. This is what impressed our philosophical ancestors, such as Locke and Kant, who were concerned with the task of understanding the nature of science as a paradigmatic knowledge-yielding enterprise that is concerned with employing conceptual devices aimed at discovering how the world really is and the characteristic regularities of the real world. In view of the new sceptical onslaughts, such as Gettier type paradoxes and riddles of induction, this task has assumed new importance.

Epistemic Norms and the role of the Cognitive — Ecological Factors in the Acquisition of Knowledge

If one of the goals of epistemology is to account for the way science comes to have knowledge of how the world really is, then one must pay more attention to the *cognitive processes*, including the *ecological* factors, involved in acquiring this knowledge. An understanding of the nature of epistemic norms will remain incomplete as long as it is not integrated into a more comprehensive account of the cognitive and ecological factors and their interactions that form part of the functioning of the epistemic agent. Only in the recent past more attention is being paid to the role of the cognitive and ecological factors in such important issues in philosophy of science as scientific observation/experiment, formation of concepts and categorization, model building and theory change. This section attempts to provide an outline of this approach and its relevance to the formulation of epistemic norm.

Scientific Knowledge and Naturalization of Belief and Knowledge: The Interaction-Information Theoretic Account of Belief/Knowledge

The cognitive, ecological, interactive and information theoretic approach, advocated, among others, by Goldman,[46] Dretske,[47] Barwise and Perry,[48] J.R. Anderson,[49] J.J. Gibson,[50] may be construed as denial of (iii) along with the justificatory account of knowledge (involving the base statement and inductive inferential view).

46. A. Goldman, *Epistemology and Cognition.*

47. F. Dretske, *Knowledge and the Flow of Information*, Cambridge, Mass.: MIT Press/Bradford, 1981.

48. J. Barwise, and J. Perry, *Situations and Attitudes*, Cambridge, Mass.: MIT Press, 1983.

49. J.R. Anderson, *The Adaptive Character of Thought*, Hillsdale, NJ: Erlbaum, 1990. See also J.R. Anderson, "The Adaptive Nature of Human Categorization", *Psychological Review*, 98 (1991), pp. 409-29. J.R. Anderson, "Is human cognition adaptive?" *Behavioral and Brain Sciences*, 14 (1991), pp. 471-517.

50. J.J. Gibson, *The Perception of the Visual World*, Boston: Houghton Mifflin, 1950. See also J.J. Gibson, *The Senses Considered as Perceptual System*, Boston: Houghton Mifflin, 1966. J.J. Gibson, *The Ecological Approach to Visual Perception*, Boston: Houghton Mifflin, 1979.

The rejection of (iii), however, will lead to one desirable consequence, viz. the blocking the possibilities of the Gettier type paradoxes by providing a description of the cognitive ecological *circumstances of acquiring knowledge* such that *one would not believe that p, unless "p" is true*. This amounts to being in a cognitive state in which *evidential justification* and *the belief that p* are not separated, rather one *involves* the other in terms of causal or nomic regularities/constraints such that it would be physically or causally impossible to be in that cognitive state of belief without the belief being true.

Epistemic acts, such as observation, according to this view, is said to involve successive type of states or situations where *information flows* from one type of state or situation to another. Thus, in observation the information that "x is P," where x and P are features of the physical world is carried from the object to the sense organs or receptors through a process of interaction between the two. This interaction relation leading to *information flow* is based on a *nomic regularity/constraint* holding between state or situation type where one *involves* the other, e.g. the fact that the X-ray has such and such a pattern *carries information about/involves/indicates* that Jackie has a broken leg.

DRETSKE'S DEFINITION OF BELIEF

According to Dretske, belief, the thinking that something is the case, is characterized in terms of

(i) the instantiation of structures (presumably neural) that have, through learning, acquired a certain *information-carrying role*, and

(ii) occupy an *executive* office in a system's functional organization and determines the system's causal influence on *output*.

DRETSKE'S DEFINITION OF KNOWLEDGE

Dretske offers an alternative model of knowledge as information-caused belief, which is intended to avoid the problems the justificatory accounts of knowledge face.

For him S knows that x is F amounts to saying that S's belief that x is F is caused (or causally sustained) by the information that x is F.[51]

THE ISSUE OF APPROPRIATE FRAMEWORK FOR REPRESENTING INFORMATION

The dominating view on representing information, belief and knowledge has been variously called the *language centred, propositional or symbolic/syntactic view*. In the context of Cognitive Science and AI, this view has been endorsed by Fodor,[52] Pylyshyn,[53] and Newell and Simon.[54] The theses of *language of thought* or *mentalese* and *physical symbol system hypothesis* forcefully articulated the *language centred, propositional* or *symbolic syntactic view* on representing information, belief and knowledge. This view is faced with several difficulties:

(i) *The symbol grounding problem*: How can the meanings of the meaningless symbol tokens, manipulated solely on the basis of their arbitrary shapes be grounded or connected up with the world in the right way?

(ii) *The frame problem*: Assuming that an intelligent agent is capable of planning and problem solving and given the fact that she is acting, can we specify in symbolic

---

51. F. Dretske, *Knowledge and the Flow of Information*, p. 86.

52. J.A. Fodor, *Representations*, Cambridge, Mass.: MIT Press, 1981.

53. Z. Pylyshyn, *Computation and Cognition*, Cambridge, Mass.: Bradford/MIT Press, 1984.

54. A. Newell and H. Simon, "Computer Science as Empirical Inquiry: Symbols and Search", *CACM*, 19 (1976), pp. 113-16.

formalism what changes and what remains constant in the particular domain? Classical symbolic systems are monotonic, whereas planning and problem solving invariably involve new experience and change.

(iii) *The problem of induction*: Mere symbolic representation does not lend itself easily to model judgements of *similarity* and to identify projectible predicates that denote *natural kinds*. Both Goodman and Quine taught us that these issues are intimately related to the problem of induction.

The basic problem with systems using propositional or syntactic symbolic representation is that they are disjoint from non-symbolic system (i.e. the world). Hence, there must be some point where the information from the world is presented in symbolic form.

TWO IMPORTANT ISSUES

There are two important issues here:

(i) How a system using propositional or symbolic representation takes in information from a non-symbolic world?

(ii) How does a cognitive system distinguish and grasp non-logically the predicates that denote *natural kinds* or *natural properties*, which represent things that exist in reality and also make induction possible?

Conceptual Space Approach as a Framework for Representing Information: An Alternative Epistemological Framework

Many cognitive scientists[55] claim that symbolic and cognitive structures are perceptually grounded. This, however, involves

55. B. Johnson, *The Body in the Mind: The Bodily Basis of Cognition*, Chicago: University of Chicago Press, 1987.

an analysis of (a) what role the information at the sub-symbolic level plays in the explanation of various epistemic, knowledge acquisitive and cognitive acts, such as observation, and (b) how (a) leads to formation of concepts and categories, which finally give rise to formulation of knowledge at the linguistic level.

J.J. Gibson[56] and David Marr[57] (especially while dwelling on his first two modules, viz. Primal Sketch and 2 ½ D) provided an analysis of (a) mentioned above with specific reference to visual information.

Gibson offers an analysis of *optical information* in which appearance in the sense of *the way things are* is directly given in *visual information* and not inferred. Therefore, his view is sometimes called *Theory of Direct Perception (TDP)*. TDP comprises two separate investigations:

- one involves examining the perceiving organism, and
- the other concentrates on the *what* of perception, examining the visible world external to the organism.

Gibson terms the latter inquiry *Ecological optics*, which involves finding environmental properties that can be uniquely and invariantly specified in the *structure of the reflected ambient light* in the form of an *optic array*. The structure of the reflected ambient light depends upon the structure of the surface of the perceived object. Unlike the radiant light from a light source, the structure of the reflected ambient light is responsible for the perception of distance, depth, motion, etc. For Gibson the structure of the reflected ambient light does not carry/convey information, *it is* information. The job of the perceptual system is merely to *pick up* the information by orienting, adjusting, resonating and tuning to the light input.

---

56. See note 50.

57. D. Marr, *Vision*, San Francisco: Freeman, 1982.

What aspects of structure in the light that are *available* in the light itself and picked up by the eye are potentially informative for perception? For Gibson there are four aspects:

1. The *gradient* structure in the light could provide direct information for various aspects of the visual world, such as distance. A gradient refers to the rate at which some dimension of the optical stimulus, e.g. the size of the angle subtended by an object, varies with respect to some other dimension of the optical stimulus, such as near and far.

2. Continuous gradient or *optical flow* with different gradient for each different degree of slant, and

3. *Invariants*: perception is dependent on pick-up not only of static structure in the light, but of structural *invariances* across changing views. The invariants refer to the properties of figures that are left unchanged by the (geometric) transformation and show that optical information can determine alone the perception of the constant or persistent properties of objects in the face of continually changing sensations, e.g. invariants of optical structure under the change of the point of observation.

4. *Affordance* or the resources of the environment-organism interaction, such as invariances, that enables the organism to survive and evolve.

Thus Gibson treated the problem of perception as that of recovering from sensory information *valid properties* of the external world.

However, as Marr points out, Gibson's analysis is incomplete. Gibson failed to realize the fact that the detection of *physical invariants*, like image surfaces, is exactly and precisely an information-processing and cognitive problem.

Secondly, Gibson vastly underrated the sheer *difficulties of such detection.*

Marr, then, goes on to provide an elaborate analysis of how the information provided at the Primal Sketch and 2½D levels lead to identification and categorization of 3-D objects as we ordinarily recognize them. It is at this level the analyses of Marr and Gardenfors appear to be complementary to each other and are relevant to (b) mentioned above.

GARDENFORS' NOTION OF CONCEPTUAL SPACE

Peter Gardenfors[58] offers an answer to the second issue by suggesting a hybrid three tier cognitive system with (a) sub-conceptual, (b) conceptual and (c) higher symbolic level. For the traditional empiricist thinkers, such as Locke and Hume, the bridge between the external world and mind was provided by *impressions, ideas* and *associations* between perceptions and *not* propositions.

Marr also accepts a three level analysis in which sub-conceptual level of *primal sketch* and 2½ D plays an important role in modelling perceptual knowledge.

Gardenfors introduces a non-linguistic and non-logical way of representing information and knowledge in terms of his theory of *Conceptual Space* in which the objects of the representation no longer form a language or have even a propositional structure. Gardenfors treats his knowledge representation framework as "cognitive" and bases it on analogical representation. He treats this as an alternative to

58. P. Gardenfors, "Induction, Conceptual Spaces and AI", *Philosophy of Science*, 57 (1990), pp. 78-95. See also P. Gardenfors (1991) "Framework for Properties: Possible World vs. Conceptual Spaces", *Acta Philosphica Fennica*, 49 , pp. 383-407. P. Gardenfors, "Three Levels of Inductive iInference" *Lund University Cognitive Studies* 9, (1992).

linguistic, propositional or Fregean representation and shows its drawbacks.

For Gardenfors *conceptual space* is "a cognitive entity," which he claims is "ontologically prior to any form of language." A conceptual space is a set of pre-linguistic *quality dimensions* that are closely connected to what is produced by our sensory receptors. Typical example of quality dimensions is colour, length, weight, temperature, time, etc. which represent various *qualities of objects* by assigning *properties* to them and specifying *relations* between them. Each quality dimension has a *geometrical, topological* or *metrical* or just *an ordering structure,* rather than syntactic or logical structure, as in symbolic models, or associations, as in connectionist model.

For example, with respect to colour there are *three* quality dimensions, such as hue, brightness, and saturation. The hue dimension has the topology of a circle (the colour circle), which enables us to talk about phenomenologically complementary colours that can be represented opposite to each other on the colour circle, such as green and red. On the other hand, the brightness dimension (i.e. white to black), and the saturation dimension (i.e. color intensity), each have a linear structure and is isomorphic to an interval on the real number line. Thus these three dimensions together, one with circular and the two others with linear structure make up the conceptual space for colour and can be represented in terms of the so-called *colour spindle.*

Gardenfors' quality dimensions are psychological dimensions and not scientific or theoretical ones. For example, he distinguishes the psychological interpretation of hue in terms of the colour wheel and the scientific or theoretical interpretation in terms of wavelengths of light. However, for him the quality dimension precedes and provides the basis for symbolic and conceptual representation.

Langacker uses the notion of domain, which is similar to Gardenfors' conceptual space:

> What occupies the lowest level in conceptual hierarchies? I am neutral in regard to the possible existence of conceptual primitives. It is however necessary to posit a number of "basic domains," that is, cognitively irreducible representational spaces or fields of conceptual potential. Among these basic domains are the experience of time and our capacity for dealing with two- and three-dimensional spatial configurations. There are basic domains associated with various senses: colour space (an array of possible colour sensations), co-ordinated with the extension of the visual field; the pitch scale; a range of possible temperature sensations (co-ordinated with positions on the body); and so on. Emotive domains must also be assumed. It is possible that certain linguistic predications are characterized solely in relation to one or more basic domains, for example, time for BEFORE, colour space for RED, or time and the pitch scale for BEEP. However, most expressions pertain to higher level of conceptual organization and presuppose non-basic domains for their semantic characterization.[59]

How does Gardenfors notion of *conceptual space* grasp non-logically the predicates that denote *natural kinds* or *natural properties*, which represent things that exist in reality and also make induction possible and distinguish them from other properties?

HOW ARE THE BASIC PROJECTIBLE PREDICATES ESTABLISHED?

Gardenfors makes a constructive use of his notion of *conceptual space* and uses it as a basis for formulating a new criterion of what a property is. He goes on to represent a *property* as a

59. R.W. Langacker, *Foundations of Cognitive Grammar*, vol. 1, Stanford: Stanford University Press, 1987, p. 5.

region in a conceptual space. And defines a *natural property* as representing a space which is *convex* in the following sense: if a pair of points a and b are in the region, then all points between a and b are also in the region.

This definition of *natural properties* makes them *perceptually grounded,* i.e. for these properties there is a direct link between the *convex region of the conceptual space* with its quality dimensions and *perception.*

According to Gardenfors the predicates *green* and *grue* differ from each other, because *green* designates a property that can be represented in a convex region, while *grue* cannot be represented by a convex region, because the predicate *grue* involves both colour as well as time dimensions (i.e. an object has the property *grue,* if it is examined before time t and determined to be green, or it is not examined before time t and it is blue).

Natural Properties, Similarities and Induction

Gardenfors contends that the *projectible* predicates are the predicates, which designate natural properties and that only these predicates make inductive inferences possible. He also goes on to define a relation of comparative *similarity* in terms of a conceptual space by letting two objects count as "more similar to each other the closer their set of properties is located in the underlying conceptual space."

Gardenfors claims that humans generally agree as to which properties are the projectible ones. This suggests that humans have *close to identical* psychological conceptual spaces. The evolutionary theory and natural selection explain why "our way of identifying natural properties accords so well with the external world as to make our inductions tend to come out right" and demonstrate that our inductive capacities are dependent on the "ecological circumstances under which they have evolved."

A Cognitive Account of Properties and Concepts

For Gardenfors a conceptual space is a set of pre-linguistic *quality dimensions with a geometrical structure*. The motivation for distinguishing several quality dimensions has to do with the fact that an object can be assigned a property *independently* of other properties, although within the same quality dimension assignment of properties cannot be done independently of each other. For example, an object can be assigned the property of weight, say, one kilogramme, independently of its temperature or colour. Hence, *different* quality dimensions are *separable* and *incommensurable*. However, there are quality dimensions that are *integral,* i.e. an object cannot be assigned a value in the same dimension, say, hue dimension, without assigning a brightness value, say, assigning the property *green* independently of being assigned *red*. He defines a *domain* to be a set of *integral* quality dimensions, such as hue, brightness and saturation that are *separable* from other quality dimensions, such as taste or sound.

Gardenfors uses the distinction between *separable* and *integral* quality dimensions and his notion of *domain* in order to define *properties* and *concepts*. A *property* is defined in terms of a single quality dimension or a finite number of *integrated* dimensions forming a *domain*. In contrast, a *concept* can be construed as *several separable* quality dimensions or domains or conceptual sub-spaces. Hence, from this point of view, *properties* can be seen as a special case of *concepts*.

Gardenfors complains that the distinction between *properties* and *concepts* is obliterated in traditional symbolic language as both are represented in terms of *predicates*, which collapse the distinction between *adjectives, nouns,* and *verbs* and club the *information content* implicit in these different grammatical categories in a natural language.

CATEGORIZATION, CONCEPT REPRESENTATION, AND CONCEPT DYNAMICS

A *concept* was taken to represent multiple domains where each domain corresponds to a quality dimension or region. For example, the concept *apple* is associated with such domains as colour, shape, texture and taste. For the concept *apple* the colour domain corresponds to red-yellow-green, the shape to roundish (cycloid), texture to smoothness, and taste to the region of the sweet and sour dimensions.

Since there are several domains involved in a concept, it is often observed that certain domains are given priority and prominence than others. The *prominence value* given to a particular domain is dependent on such contextual factors as world knowledge or interest of the agent. This results in the distinction between properties into *essential* and *incidental* or *core* and *peripheral*. The former distinction of properties leads to *Essentialism* and the latter to a more favoured theory known as *Prototype Theory*. The *Prototype Theory* holds that within a category of objects, like those instantiating a property or concept, certain members are judged to be more *representative* than others.

Categorization, i.e. the act of determining which objects belong to or does not belong to class, in terms of the classical notion of property, viz. *essentialism* or *intensional semantics,* fails to provide a clear criterion for categorization. As opposed to this Gardenfors' criterion of natural property as a convex region of a conceptual space can easily explain prototype effects, since in a convex region one can describe positions being more or less central. For example, if colour properties are represented in terms of the convex sub-sets of the colour space, the central points of these regions can be considered as the most prototypical examples of colour.

Gardenfors takes some quality dimensions of a conceptual space S as a Euclidean space so that for every point p in the space one can measure the distance dE (p, pi) from p to each of the pi, satisfying the conditions of *distance function.* A space with a distance function is called a *metric space.* A Euclidean metric then is defined on the subspace that is subject to categorization. The line between pi and pj is called a bisector. Each bisector divides the plane into two halves.

Gardenfors then appeals to a result, known as *Voronoi tessellation,* in order to represent categorization in terms of his definition of property as a convex region. The result claims that the *Voronoi tessellation* based on an Euclidean metric always results in a *partitioning* of the space into *convex* regions with respect to the point p_i.

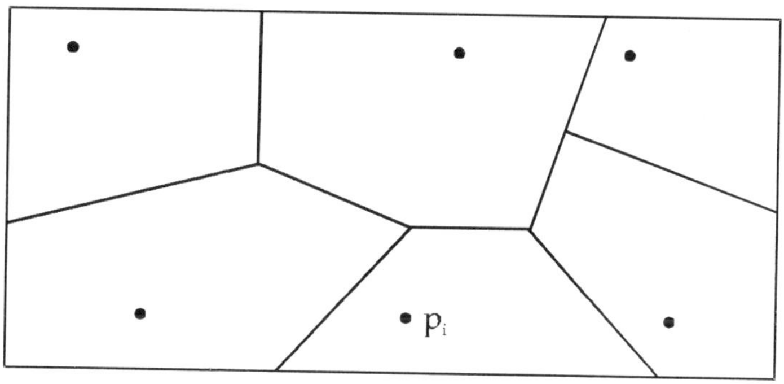

The dynamics of concepts and categorization, e.g. change from general category to subordinate or change in relative prominence resulting in re-categorization, can be explained with the aid of the *Voronoi tessellation* by changing the relative distance between the points or shifting the borders as a result of a change of the scale of the x or y-axis. Thus, the concept dynamics represented in this way may lead to a concept-based form of non-monotonic reasoning.

III

An Alternative Concept-based Approach to Belief Revision and Theory Change

Many Cognitive Scientists have suggested that knowledge could be represented non-propositionally in terms of concepts and investigated the nature and psychological functions of concepts.[60] Recent works on concepts, categorization and conceptual change point to the fact that the ability to categorize depends upon fairly complex knowledge, which may be encoded more like a pattern and is much more than mere propositions.[61]

Although the propositional representation of beliefs is a more abstract one, it does not reflect the beliefs and ideas

---

60. E. Rosch, "Cognitive Representation of Semantic Categories", *Journal of Experimental Psychology : General,* 104 (1975), pp. 192-233. See also E.A. Smith, and D.L. Medin, *Categories and Concepts,* Cambridge, Mass.: Harvard University Press, 1981. D.N. Osherson, and E.E. Smith (1981) "On the Adequacy of Prototype Theory as a Theory of Concepts", *Cognition,* 12 (1981), pp. 299-318. E. Smith (1989) "Concepts and Induction" in M. Posner (ed.), *Foundations of Cognitive Science,* Cambridge, Mass.: MIT Press, 1981, pp. 501-26.

61. J.R. Anderson, *The Architecture of Cognition,* Cambridge, Mass.: Harvard University Press, 1983. See also, J.F. Sowa, *Conceptual Structures,* Addison-Wesley, 1984. D.S. Touretsky, *The Mathematics of Inheritance Systems,* Los Altos, CA: Morgan Kufmann, 1986. H. Margolis, *Patterns, Thinking, and Cognition : A Theory of Judgement,* Chicago: University of Chicago Press, 1987. W. Bechtel, and A. Abrahamsen, "Beyond the Exclusively Propositional Era", in J.H. Fetzer (ed.), *Epistemology and Cognition,* Dordrecht: Kluwer, 1991. P. Thagard, (1992) *Conceptual Revolutions,* Princeton: Princeton University Press, 1992.

involved in making a particular proposition. Thagard (1986) claims that whenever concept changes, it does so by virtue of changes in the beliefs that employ the concept.[62] A concept can be thought upon as mental structure analogous to data structures in computers. Moreover concepts can help us in categorization, learning, memory, deductive inference, explanation, and problem solving. An understanding of conceptual development in science and everyday life is an essential part of epistemology. Merely attending to belief revision will miss crucial aspects of epistemic change.

However, an analysis of epistemic change in terms of conceptual change involves:

(i) *Representation of concepts*: Unlike Gardenfors' attempt to represent concepts using geometric ideas of conceptual space, the Inheritance Conceptual Network view uses the idea of points/nodes and arcs in order to represent concepts and their relationships.

An understanding of concept-based cognitive change/ theory revision involves much more than a view of the nature of isolated concepts.

(ii) *Representation of the complex structure of concepts*: We need to see how concepts can fit together into conceptual system and what is involved in the revision or replacement of such systems. This structure in the form of a net or tree and the associated semantic properties: A semantic-net-style organization of concepts will involve a hierarchical organization of concepts in terms of *kind-hierarchies* and *part-hierarchies* linked to each other by rules represented in terms of arks representing *is-a relation*.

---

62. P. Thagard, (1992) *Conceptual Revolutions*, Princeton: Princeton University Press, 1992.

(iii) Thagard provides a taxonomy of the conceptual changes, which involves belief revision based on other operations over and above addition and deletion of beliefs. The taxonomy includes:

Thagard lists some of the important operations that can be performed on a conceptual net leading to conceptual and

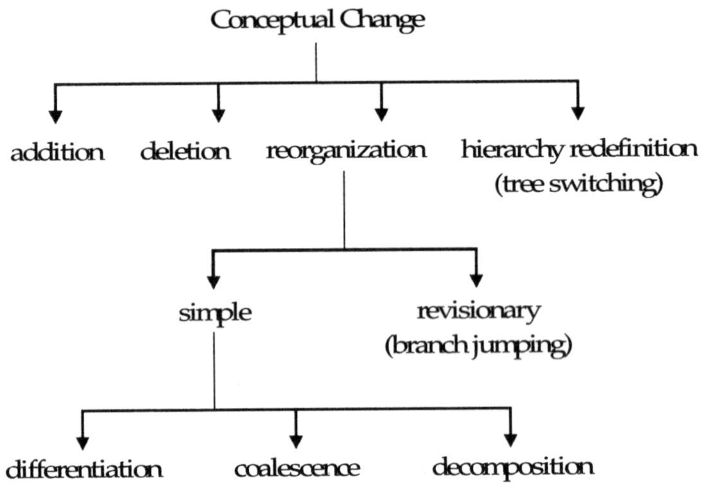

epistemic change. They are :

1. Adding new instances,
2. Adding new rule,
3. Adding new part-relation,
4. Adding new kind-relation,
5. Adding a new concept,
6. Collapsing part of a kind-hierarchy, i.e. abandoning a previous distinction,
7. Reorganizing hierarchies by branch jumping, i.e. shifting a concept from one branch of a hierarchical

tree to another, e.g. Darwin's re-categorization of humans as a kind of animal, and

8. Changing the organizing principle of a hierarchical tree, i.e. tree switching.

Thagard undertakes detailed case studies, e.g. the chemical revolution, to illustrate how radical theory change can be understood within his framework of conceptual change.

Conclusion

The Received View on Epistemology based on the notions of conditions of belief, truth and justification is firmly committed to the propositional and representational approach. The developments in Cognitive Science provide evidence that goes on to show that the information about the way the world is can be obtained and revised in terms of frameworks that are non-propositional. We have reviewed such approaches and found that all the approaches take sub-symbolic grounding of concepts as the basis for "knowledge representation" and revision. These developments should be taken into consideration in any viable epistemology.

PART II

Cognition and Science

9

Explanation, Causation and Physical Necessity

From the Standpoint of Internal Realism

Chhanda Gupta

I

> A prime objective of science is to explain why things happen in the way they do. Its aim is to provide explanations of events and their recurrence in terms of laws — of laws by theories, and of such theories by still more comprehensive and fundamental theories.

THUS, one may wish to explain why, for instance, in a particular case a certain amount of sodium in chemical combination with a certain amount of chlorine form the compound common salt. Why again does this happen recurrently, one could further inquire. Then, to explain both, the particular event and its recurrence, one may want to subsume them under a lawful empirical generalization like: whenever sodium is chemically combined with chlorine in a fixed proportion, the compound salt is formed. This, as we know, was long regarded as the standard way of explaining pace Hempel. On this view, we have at the most basic level particular empirical events including those that recur. At the next level these events are subsumed under empirical laws of the kind mentioned. At the next higher level these laws are sought to be explained by theories, and the theories themselves by still more comprehensive theories. So, while the valency theory explains why atoms of certain elements combine in definite proportions regularly, lawfully, the valency theory itself can be explained in terms of more fundamental theories of atomic physics. The reason why atoms of two or more elements combine chemically in definite proportions, for example, is to be found in the combining capacity or valence of an atom. And this is determined by the number of electrons which an atom has in its outermost shell and which it tends to lose or gain in combining with other atoms.

II

Kinds of Explanation

If this be the explanatory story level by level, then explanatory knowledge might seem to be based on our descriptive knowledge of the natural world. We can explain, it seems, given this descriptive view, by simply describing how things actually are. But mere description, even if it is of the most complete sort, does not seem to answer the question "why", which is what we typically raise in the context of explanation. In that context, we feel perplexed if we do not understand why what actually happens, has to happen the way it does, invariantly and exceptionlessly. Why-questions stem from this sense of perplexity, and explanations are supposed to be answers to such why-questions.

For this reason, many philosophers favour a different account of explanation, a modal account, in which explanatory knowledge exceeds the bounds of descriptive knowledge. To explain an event it is not sufficient to assert merely that events like it do happen, or even happen exceptionlessly as a matter of fact. It has to be shown why this event, or events like it *had to happen*, and why even the possibility of an exception to such events is ruled out. In other words, just a statement of the assertoric form expressing a *de facto* regular connection between two types of events does not explain. Statements which can explain must be expressions of a stronger modality. So a mere description of the form: "All F's are G's" will not do. A stronger modal dimension must be added to the description. When this is done, "All F's are G's" does not merely state:

1. All past F's were G's;
2. All present F's are G's;
3. All future F's will be G's; as a matter of fact.

It assumes instead the form of a strictly lawlike or nomological generalization: "All F's must be G's," which rules out a situation where something which is F is not G. This stronger modal statement is prohibitive in the sense that it prohibits a certain situation as impossible (the situation of an F not being G for example). More precisely and specifically, it excludes certain situations as physically impossible. Explanatory knowledge, according to those who advocate this account of explanation, is knowledge of what is *physically necessary* and *physically impossible*.

There is another widely prevalent view of explanation. To explain some phenomenon is to say how it was caused. To explain the formation of a chemical compound, to take our old example, is to state what brought it about. Though this sounds simple, the notion of causation has been a worry for many philosophers, particularly for those who have an empiricist bent of mind. Empiricists find the notion problematic inasmuch as it is believed to involve "physical necessity."

Given the necessitation view of causation, an event to qualify as a cause must necessitate, or physically necessitate, the event which is the effect. This means that if the cause occurs, the effect must necessarily follow. It is physically impossible that the cause occurs and the effect does not. It is impossible, for instance, that there be a transfer of an electron (and so of a negative charge) from a sodium atom to the outermost shell of a chlorine atom which is incomplete with electrons, and yet there is no bonding between them.

However, if causation is thus understood to involve necessity, then the explanation given in terms of causation will tend to merge with the modal model of explanation. And this creates problems for the empiricists who have gone to the extent of demanding the complete extrusion of the word "cause" from the philosophical vocabulary, for they are uncomfortable about the notion of observationally inaccessible "necessity" and its kindred notions like "causal power" and "disposition." Yet there are many others who believe that causality, notwithstanding the necessity it involves, lies at the heart of our attempt to understand nature. The invariance and orderliness of observed regularities at the surface level, they point out, are due to necessary causal connections of a deeper level, and these connections in their turn are grounded in deeper fundamental features of things that are not accessible to observation. But what are we to understand by these fundamental features which constitute objective necessities — necessities *'de re'* as they are called, being necessities in *res* or things?

1. One answer to the question dates back to Aristotle's *Posterior Analytics,* where "necessity" is explained in terms of "essences," and the concept of "essence" is understood with reference to the notion of

"substance." Essence is equated with substance, which is being or existence. Essence is the very nature of such being or a thing, which makes it what it is and gives it its identity. Minus this essence, the thing is pure being or existence, which according to Aristotle, is impossible. What, however, constitutes the essence of a thing, or why and how does a thing come to be what it is? The answer is given in terms of his doctrine of four causes, and it is the final cause which is said to constitute the essence of a thing. This is suggested by the phrase he uses in this context, namely, *to ti en einai atoi,* which means: the what is was for so-and-so to be.[1] Essence thus is what anything is or has to be in order to be so and so. Both the nature of a thing and what it will be or do are determined by essence.

2. A different answer is given by some rationalists. Objective necessity is believed to be an unconditional necessity, and this is believed to emanate from a supernatural power.[2] The necessity enshrined in laws which govern the course of things and events which are causally related, according to these thinkers, holds unconditionally.

3. Locke's account of objective necessity does not have such abstract metaphysical and theological underpinnings. It has a scientific orientation. Influenced by the corpuscular theory, he finds the locus of objective necessity in the physical structure of insensible particles of which the material bodies are

---

1. R. Sorabji, *Necessity, Cause and Blame,* London: Gerald Duckworth, 1980, p. 191.

2. M. Bunge, *Causality: the Place of Causal Principle in Modern Science,* Cambridge, Mass.: Harvard University Press, 1959, p. 40.

made, and which determine all their observable qualities. Locke identified what he calls "real essence" with this physical structure of the insensible particles of substances. The real essence of gold, for example, is the internal constitution or structure which its insensible particles have, but of which men are ignorant. The "nominal essences" of things by contrast are ideas of those characteristics which are readily detectable, and by virtue of which we can recognize and classify objects into various sorts. So, even if Locke was influenced by science, the "real essences" he admits, are by strictly empiricist standard we "know not what," a concept which scientists are likely to dismiss as too obscure.

The upshot of the three accounts then is that even though causality and causal necessity seem to be necessary for scientific explanation, the way they are conceived in these accounts is irreducibly metaphysical. It is doubtful whether explanations given in terms of such metaphysical notions of necessity can be legitimated in science. But then are we to quit talking about necessity altogether and be contented with a purely descriptive model of explanation? Could we not demystify instead the notion of necessity by eliminating dubious references to "ends" or 'purposes" (final causes) for which things are brought into being, and also to supernatural agencies and hidden constitutions of insensible particles? My concern here is to show that we can. We may have explanations which are *modal*, and which do retain the notion of *necessity*. But that necessity is not metaphysical in the obscure transcendent senses noted above. It is *physical necessity;* and explanations given in terms of such necessity can be said to be based on our descriptive knowledge of the natural world.

III

Hume on Causation

Empiricists by and large will not endorse this view. The laws in terms of which one attempts to provide scientific explanations in their opinion should assert no more than constant conjunctions. Scientific laws need not be supposed to express an extra element, that is, an additional necessary connection over and above factual uniformity. David Hume was right, they think, in maintaining that objectively there is no difference between "universals of law" and "universals of fact." He was right when he urged that it is only a psychological fact about the way in which our mind works that causes us to ascribe necessity to such laws. The idea of necessary connection is derived from our experience of constant conjunction of impressions and not from anything in nature over and above constant conjunction. Echoes of this line of thinking are heard in the works of Braithwaite, along with other staunch empiricists like Ernst Mach and Karl Pearson, and also in more recent works like those of Bas van Frassen and Nancy Cartwright. However, did the "Past Master" really hold the belief which is being attributed to him? There are several interpretations of his view on causation and necessity, of which a few may be mentioned.

1. A *skeptical interpretation* is given by Green, Thomas Reid, Beattie, James Mill and John Stuart Mill. If causation involves "power" and necessary "connexion" and if these terms have no impression-derived meaning, then on semantical and epistemic grounds Hume must have drawn the negative metaphysical conclusion that there is no objective necessary causal connection.

   However, even if Hume's skepticism does make a

negative *epistemic* claim that we cannot know causal power and necessity, it does not make the negative *metaphysical* claim that there is no such thing as objective causal connection.

2. Again even if Hume's non-committal skepticism forbids one to draw metaphysical conclusions, both negative and positive, some commentators (Richard Taylor, Basson and Robinson) attribute to him the metaphysical view that causality is *pure regularity* — a view which eliminates completely the idea of necessity.
3. Norman Kemp Smith considers this pure regularity account to be grossly mistaken. He offers instead a *"modified necessity"* interpretation.[3] Hume is no supporter of the pure uniformity or regularity view, for he distinguishes mere sequence from a causal sequence, the differentia being the idea of necessitation, which again is subjective or psychological. Constant conjunction of observed items generates a habit or custom whereby the mind is led from present impressions or ideas to their usual attendants, and by doing so generates further a feeling of necessitated transition, that is, a feeling of being determined. Necessity here is understood as a determination of thought.
4. In addition to this modified necessity account, Kemp Smith attributes to Hume a *"power realist"* view of causation.[4] He sees Hume not as a regularity realist, nor merely as a modified necessity theorist, but as a power realist, who believes in objective casual necessity in nature. Such necessity is grounded in

---

3. N.K. Smith, *The Philosophy of David Hume*, London: Macmillan, 1941, pp. 91-92.

4. Ibid., p. 397.

intrinsic properties and powers which things have, and by virtue of possessing which they bring about effects necessarily.

Since this harks back to the rationalist thesis which compares causal necessary relation with entailment, and since Hume had challenged this rationalist realist tradition, many philosophers may have misgivings about the power realist interpretation. Still a very influential line of thinkers follow Kemp Smith and defend it. Power realism does not contradict Hume's basic empiricist conviction that if words like "necessity" and "power" mean anything at all they mean only a determination of thought. For one must not forget in this context the distinction between causal necessity as it is out there in the world and our idea of such necessity. *What there* is need not be equated with *what we know.* Michael Costa, endorsing Smith's view, quotes from the *Treatise* to substantiate his point.[5] Hume was ". . . ready to allow that there may be several qualities . . . in objects, with which we are utterly unacquainted. . . ."[5] Obscurity arises only when we apply the idea which we do have of "power" or "necessity"(derived from an inner impression of reflection) to such unknown qualities in objects. Hume had no intent to deny the causal structure of the Universe independent of human thought. What he denied is our knowledge of this structure. We may be unable to ideate the hidden qualities or to "positively contentfully mean" them, as Galen Strawson says;[6] nevertheless we can refer to these objective fundamental features. Both Galen Strawson and Costa delink reference and meaning. We can refer to something even though we do not have any clear conception of its nature. The ontic admission

5. M. Costa, "Hume and Causal Realism", *Australasian Journal of Philosophy*, 67, (1989), p. 180.

6. G. Strawson, *The Secret Connexion*, Oxford: Clarendon Press, 1989.

of intrinsic qualities constituting objective necessities therefore does not collide with Hume's meaning empiricism.

This delinking of meaning and reference of the terms "causal necessity" and "power," however, would produce a yawning gap between what there is and what we know. Hume himself seems to have felt the need of narrowing the gap, and claimed that there is a pre-established harmony between what there is, namely, the "course of nature" and the "succession of ideas."[7] But the harmony assumed here is between the succession of ideas and the events observed in the course of nature, not with powers or forces which govern such a course of nature. The powers and forces are wholly unknown — something with which "we are utterly unacquainted." But to speak of something of which we have no idea is to usher in just that mysterious kind of metaphysical necessity which empiricists are anxious to avoid. Hume was on to something important when he spoke of the harmony between what there is and what we know. But he could not realize this within the empiricist framework of his philosophy. To save his meaning empiricism, he had to label "power," "force," and "necessary connection" as *secret*. But to admit secret power and necessity is to deepen the gap he wished to reduce. To lessen the gap, "power," "force," and "necessity" must be demystified. And to do so, one must stop regarding them as *secret* — as something we know not what.

IV

Physical Necessity: From the Standpoint of Internal Realism

A different kind of "power realism" to which one could resort

7. D. Hume, *Enquiry Concerning Human Understanding*, Calcutta: Progressive Publishers, 1967, p. 46.

at this point is one that challenges this very idea of secrecy. It challenges the whole idea that there is a cut between objective powers, forces, and necessities with which we are "utterly unacquainted" on the one hand, and our ideas and conceptions of these on the other. This is the kind of power realism which has to be understood from the perspective of "internal realism." The basic idea of internal realism is that we can only have an adequate empirically meaningful discourse about what there is, if, whatever we are talking about is *relatable* to our thoughts, conceptions and experiences. Whatever objective features things, processes, events, and laws (including causal laws) have, cannot be intelligibly talked about without reference to the ways these appear to us. And the *ways things appear* are the *ways they really are.*

Consider an object which is warm or has a certain degree of temperature for example. Is there any sense, an internal realist would urge, in saying that it is warm by itself apart from the touching hand or has a degree of temperature by itself apart from a measuring device? Objections would be raised of course against this internalist claim.

1. Hard-core "external realists" will straightaway dismiss the claim that this relational quality warmth, which is an idea of secondary quality (in the Lockean sense) is real. It is on the contrary a mere appearance — a projection. What are real are "power" and the subvisible structure which the thing has.

   To this the internal realist's answer is: causal power and subvisible structure are real no doubt, but there is no reason to reserve reality only for these to the exclusion of qualities like warmth. These qualities too are real in spite of being relational.

2. A more vehement objection to the internal realist contention about the relativity of things and their features to our thoughts is that subvisible features, qualitative and structural, can on no account be called relational. Ideas of secondary qualities may be grudgingly called relational but not the deeper level features. Warmth — yes. But what about the mean kinetic energy of molecules? What is the atomic structure or the chemical composition of a substance, say, water to be relativized to?

To answer the question one could consider a complex set of ideas, which are central to Putnam's "internal realist" approach to the issue of physical necessity, necessity grounded in microstructure or chemical composition of substance. The main points of this approach are:

(a) *physical necessity* is to be explained in terms *"rigid designation"*;

(b) *"rigid designation"* is to be understood in terms of *substance-identity;* and

(c) *Substance-identity* should be understood in a theory-relative way.

We may try to understand these points with reference to the example of water. Water as we all know is H_2O. This chemical composition determines the identity of the substance. Now we may imagine a hypothetical situation, a "twin earth" or a "possible world," to use Putnam's and Saul Kripke's metaphors respectively, where some stuff resembles this water, displaying the same observable features which water on earth displays. Can we call it "water?" The answer will be "yes" only in case the hypothetical stuff in the "twin earth" or "possible world" is substance-identical with, or has the same microstructure as, actual water in the actual world, whether or not it displays the other observable features of earth water. It is only in this case that we can say that the substance name "water" *rigidly designates*. The substance name "water" in its rigid use refers to whatever stuff has the same microstructure as that of terrestrial samples of water. This is true whether we refer to some stuff in the actual world or to some stuff in some possible world. Since the chemical composition of the sample of actual water is H_2O, whatever stuff, actual and hypothetical, is H_2O is water. To put it differently, water is H_2O in all worlds, actual and possible. And if so, water is *necessarily* H_2O. The necessity here is explained in terms of rigid designation and rigid designation in terms of substance-identity or having the same chemical composition. So necessity here may be termed chemical necessity, or more generally *natural* or *physical necessity*.

This necessity is objective and non-epistemic. So if water is necessarily H_2O in the objective sense, then it is impossible that it is not H_2O. Like necessity this impossibility too is objective. Of course, one may conceive that water is not H_2O. But conceivability is epistemic while possibility is non-epistemic. Necessity, understood in terms of possibility/impossibility is non-epistemic. Thus regarding the chemical composition on which such necessity ultimately depends, Putnam says, it does not matter "whether we *knew* that chemical composition or not."[8] But how in that case can the chemical composition, which the stuff we call water possesses, be relatable to thought, knowledge and experience? Can necessity depending on that chemical composition be said to be *relative* to anything after all?

8. H. Putnam, "Is Water Necessarily H_2O?", in J. Conant (ed.), *Realism with a Human Face,* Cambridge Mass.: Harvard University Press, 1990, p. 59. Emphasis added. See also pp. 56-58.

To internal realists the claim concerning this *relativity* is crucial. Their special stake is to show that even fundamental features, like the chemical composition of water, can be said to be "relative" in some sense. To understand this sense and explain why in spite of being relative such features do not forfeit their claim to objectivity, we need to consider a distinction which Putnam draws between two kinds of objective non-epistemic necessity. Of these one is stronger than physical necessity. Kripke believes in this stronger variety of necessity which he calls *metaphysical necessity*. Putnam on the contrary defends the notion of *physical necessity* in contradistinction to Kripke's metaphysical necessity.[9] Both metaphysical necessity and physical necessity are non-epistemic, but the former that is the stronger, is "radically non-epistemic." Something is radically non-epistemic in the sense that it is entirely independent of our beliefs and conceptions. This is a view which most realists advance claiming that reality and truth about reality are entirely independent of our conceptions and beliefs. Davidson, recalling Putnam's characterization of transcendent realism, calls such a view of reality and truth about reality "radically non-epistemic."[10] Given this view, it is possible that all our beliefs and conceptions may be just what they are, and yet reality and truths about it may be entirely different from what we believe and conceive. Metaphysical necessity and the essential composition in which this is grounded can be said to be entirely independent of our beliefs and conceptions in this sense. Putnam does not accept this notion which is non-

---

9. H. Putnam, "Is Water Necessarily H_2O?" pp. 57-59.

10. D. Davidson, "The Structure and Content of Truth", *Journal of Philosophy*, vol. LXXXVII, no. 6 (5 June 1990), p. 298.

relational. Physical necessity, which he does accept, is non-epistemic but not radically non-epistemic. Having the same physico-chemical composition and obeying the same laws which depend on such composition are taken by him as criteria of substance-identity. These explain physical necessity and possibility. No ". . . real metaphysics is involved over and above what was already involved in taking physical possibility to be an objective notion."[11]

However, did not Kripke also think that composition determines substance-identity? Surely he did. That water is H_2O is metaphysically necessary according to him. He also maintained that this necessity is grounded in the essence of the substance, in this case, in its chemical composition. But suppose that before it was discovered to be H_2O, someone conjectured that water is something else, say XYZ. Then after the discovery that it is H_2O, what will that person say? Indeed if "water" rigidly designates only that stuff, which as a matter of objective metaphysical necessity is H_2O, then that person will have to say that the stuff he conjectured to be XYZ was not in fact XYZ, but only an epistemic counterpart of XYZ. He might have once felt certain that it was XYZ; but as Kripke emphasized necessity is metaphysical and certainty epistemic. So, what the person must say after the discovery that water is H_2O is that it is metaphysically impossible for water to be XYZ. He would therefore be unwilling to say that any logically possible situation is one in which water is XYZ.[12] But then will he be willing to say if the discovery that water is H_2O is itself a mistake? As in the case of XYZ, he will have to say that the stuff conjectured to be H_2O was not H_2O in fact but only an epistemic counterpart of H_2O, and further, he will be unwilling

11. H. Putnam, op. cit, pp. 60-61.

12. Ibid, p. 61.

to say that any logically possible situation is one in which water is H_2O. What all this shows is that our willingness to call a stuff "water" in a possible situation depends on empirical discoveries that represent states of human knowledge expressed in the form of theories. Now even though Kripke himself says that necessities in nature are discoverable a posteriori, can he also say that such necessities are metaphysical which would be just what they are in themselves irrespective of what our states of knowledge are? I think Kripke can opt for anyone of the two following positions, not both:

1. He may hold on to the radically non-epistemic notion of metaphysical necessity, which is just as it is by itself, irrespective of empirical discoveries. Empirical discoveries may force us to say that what we were willing to call "water" in a hypothetical situation, namely, XYZ or H_2O were not XYZ or H_2O in fact but only their epistemic counterparts. For what "water" rigidly designated could be very different as a matter of fact from what we claim to have discovered. Since hard facts are what the radically non-epistemic notion of metaphysical necessity demands, and empirical discoveries leave us with epistemic counterparts only, empirical discoveries have little relevance for understanding metaphysical necessity. To understand this, or what "water" rigidly designates in all possible situations, we may simply intuit that the hypothetical stuff in such situations is the same as actual water on earth. And what "same as" means here is "precisely what "=" means in logic. It means identity. Identity is a primitive logical notion."[13]

---

13. H. Putnam, op. cit, p. 66.

2. Alternatively, Kripke could maintain that necessity depends on substance-identity determined by the composition of the stuff in question. In that case we cannot understand "same as" simply in terms of the primitive logical notion of identity. To say that some stuffs, say A and B are substance-identical, one has to say they have the same composition, which Kripke did say. He also claimed that this was discoverable a posteriori. But in that case he should quit talking about metaphysical necessity, because there is something in that very notion which makes what we come to believe through empirical discoveries irrelevant. For metaphysical necessity is just that kind of a radically non-epistemic fact which may be very different from all our beliefs.

Putnam believes in physical necessity which is objective and non-epistemic in the sense of being something which is not a projection. This is different from what non-epistemic typically means when attached to metaphysical necessity. When so attached, it means something which is entirely independent of, and may be very different from, all our beliefs and conceptions held true by us at various stages of our cognitive enterprise. This is to say that it is "radically non-epistemic"; it is something transcendent which transcends all states of knowledge. In the case of physical necessity on the contrary, non-epistemic signifies that which is not an imposition or projection. It is knowable, and therefore cannot be said to transcend all states of knowledge. But it is not a projection of the knowing mind either. When it was discovered for instance that water is H_2O, no realist would say that the chemical composition was a projection of the knowing mind or of a certain theory which the mind conceives at a certain stage of inquiry. The theory does not impose the property of having this chemical composition to a stuff which in reality is something very different. The stuff really has the composition which the theory conceives it to have, just as a body really has the warmth the touching hand feels it to have. This claim to be intelligibly made however, must involve a reference to us — our perceptions, beliefs, conceptions and theories. Substance-identity, depending on microstructure must be understood in a theory-relative way, or else severing all relations with us, with our ways of viewing or conceiving it, will make it collapse into something that is wholly transcendent — "secret" — something which is "we know not what."

The claim that the stuff really has the composition which our best available theory conceives it to have, is defeasible. There is no warrant, it seems, for saying that this theory is true. The very claim that the way things appear is the way they really are, is suspect. But internal realists do not put all appearances on a par, claiming truth for each one of them. Like any fallibilist, they admit the possibility of error. One theory may be replaced by another, which through long-ranging investigation may become entrenched. It may still be argued that even erroneous conceptions get entrenched. But even if we admit that not all that appear to our theories are real, it does not follow that all are unreal either. Can fallibility be a plea for saying that all our conceptions are false? To say so is to give into total skepticism, which is incoherent.

Since total skepticism is incoherent, we have to say, it follows, that some of our beliefs are true. To establish this, Putnam and Davidson offer a transcendental argument.[14] The

14. D. Davidson, "A Coherence Theory of Truth and Knowledge", in E. Lepore (ed.), *Truth and Interpretation : Perspectives on the Philosophy of Donald Davidson*, Oxford: Basil Blackwell, 1986.

point of the argument is to show that most of our beliefs are true. But can conflicting beliefs be true? One need not say that they are. However differently things and their features are viewed by the disputants from within their conflicting scientific frameworks, they do communicate nevertheless with each other. Otherwise they would not even know in what respects they differ. To explain how this communication is possible, we have to assume that many of the beliefs held by the proponents of different theories are coherent. Short of coherence between beliefs held, they would not have been able to communicate. This is the first stage of the argument. But how is coherence itself to be rendered intelligible? The second stage of the argument is that we must assume that most of our beliefs in a coherent total set of beliefs are true. Unless the coherent beliefs are assumed to be true, the coherence of one's own beliefs and that of his beliefs with others' beliefs will not be possible. The coherence is due to the fact that objects and events about which we have these beliefs are really what we believe or take them to be.

This argument may be supplemented by another, an explanationist argument, which strengthens the claim that a stuff really has the composition which a theory conceives it to have. After decades of painstaking inquiry, proponents who had initially advanced conflicting theories, may eventually agree that water is H_2O. The beliefs entertained by some theorists about its chemical composition could turn out to cohere with those accepted by others, who at one point had mistakenly supposed that it was something else, say, XYZ. These shared coherent beliefs gradually become settled, especially when the investigators experience exactly what they expect and predict on numerous occasions, on the basis of their shared beliefs about the microstructure of the substance in question. When the stuff behaves exactly in the way expected, one can claim perhaps that their beliefs are not only justified, but also true. This is the best explanation of the successful predictions made on the basis of such beliefs. If we do not acknowledge the truth of the beliefs, success will be a miracle. Realism is the only philosophy that does not make the success of science a miracle.

10

The Observational-Theoretical Distinction Revisited

Kausik Bhattacharya

One of the difficulties of the positivist criterion of cognitive meaning is that the criterion appears to be stipulative. It appears that the positivist was not really trying to establish a criterion of cognitive significance; rather he was trying to find out a criterion of demarcation between science and metaphysics with the intention to stipulate it as a criterion of meaningfulness in order to reject metaphysics as meaningless. Unless the positivist establishes that the criterion is a criterion of cognitive significance in addition to a criterion of demarcation between science and metaphysics, the initial aim of the positivist of rejecting metaphysics as meaningless and of eliminating metaphysical entities from philosophy and science would be vitiated. Still logical positivism can survive as a philosophy of science if it can offer an adequate criterion of demarcation between science and metaphysics. Unfortunately, each formulation of the positivist criterion of verifiability either excludes as meaningless much that the positivist did not wish to exclude, or include as meaningful the most blatant non-sense. But the translatability criterion formulated by Carnap and Hempel appears to be more prospective at least as a criterion of demarcation.[1] Thus, logical positivism survives as a philosophy of science long after its death as a general epistemology.

The positivist assumed that the statements of science, unlike the statements of metaphysics, are translatable into an empiricist language whose non-logical terms are observational terms referring to directly observable physical entities or directly observable properties of physical entities. The theoretical terms are given an explicit definition by means of observation terms. This programme not only aims at providing a demarcation between science and non-science but also aims at providing the structure of scientific theories exhibiting how the abstract concepts and laws of science are ultimately related to experience and derived from empirical generalizations about observables.

The positivist philosophy of science is based on the distinction between observational and theoretical terms, and many post-positivist philosophers like Achinstein, Putnam and Popper have rejected this distinction. The main charge is that such a distinction cannot be drawn on the basis of ordinary scientific usage of non-logical terms. Thus, a so-called observational term like "red" is not really an observational term in the sense that it refers to only directly observable properties. It refers to dispositional and causal properties as well and stands for causal descriptions which transcend experience.

Popper holds the view that all universal terms are dispositional — they are used to characterize the law-like

---

1. For a precise account of the translatability criterion see C.G. Hempel, "Empiricist Criteria of Cognitive Significance: Problems and Changes", in A.P. Martinich (ed.), *The Philosophy of Language*, New York: Oxford University Press, 2001, pp. 34-46.

behaviour of certain things.[2] That is, universal terms entail universal laws. Generally we think that a term like "soluble" or "breakable" is dispositional, whereas a term like "red" or "white" is purely observational. But Popper argues that the term "red" is also dispositional. We say of a surface that it is red if it has the disposition to reflect a certain kind of light, and if it looks red in a certain situation. Again, it may be added that the concept of red involves the concept of light which, in turn, involves the concept of wave. As the concept of wave is a theoretical concept, the concept of red involves a theoretical concept. Thus, all terms are theoretical to some degree, though some are more theoretical than others.

A so-called observational term transcends experience in two ways. First, in addition to one or more directly observable properties it refers to dispositional or causal properties which might remain non-occurrent when the term is used in a context. Secondly, it entails universal laws, and universal laws transcend experience only because they are universal and thus transcend any finite number of their observable instances.

Putnam also shares the view that universal terms refer to universal laws. From the idea of cluster concept he derives the idea of law-cluster concept and maintains that universal terms are associated with the law-cluster concepts.[3]

The charge against the observational-theoretical distinction is that such a distinction cannot be drawn on the basis of ordinary usage of non-logical terms in natural scientific language, but not that we cannot construct an artificial

---

2. K.R. Popper, *The Logic of Scientific Discovery*, London: Hutchinson & Co., 1975, pp. 424-25.

3. Putnam, "The Analytic and the Synthetic", in H. Feigel & G. Maxwell (eds.), *Minnesota Studies in the Philosophy of Science*, vol. III, Minneapolis: Univ. of Minnesota Press, 1962, pp. 378-79.

language imposing an artificial distinction on the non-logical terms. Of course, we can propose an artificial language in which we may, for instance, employ "red_o" to refer to observable occurrences of the property *red*, and "red_t" to refer to non-observable or dispositional or causal properties. But the philosophical significance of drawing the distinction in such an artificial way can be and has been questioned.[4] The observational-theoretical distinction fails to guarantee the cognitive significance of theoretical terms. Still such a distinction would be significant in understanding the development of science if the development proceeded upwards from empirical generalizations about observables to progressively more abstract generalizations about non-observables. But it has been argued that this is not the true picture of the development of science. Thus, it appears that an artificial distinction between observational and theoretical terms lacks any philosophical significance.

However, the rejection of observational-theoretical distinction invites a difficulty in Popper's realistic framework. The ground of rejecting the distinction is that a so-called observational term stands for causal and dispositional properties and entails universal laws which transcend observation. But to formulate a universal law we need universal terms. Now, if all universal terms entail universal laws and we need universal terms to formulate universal laws, then we either get an infinite sequence of universal terms and universal laws or get involved in circularity. Since a language contains a finite number of universal terms, circularity seems to be the inevitable consequence of rejecting the observational-theoretical distinction on the ground that all universal terms

4. For an elaborate discussion see F. Suppe, "Criticisms of the Received View", in F. Suppe (ed.), *The Structure of Scientific Theories,* Urbana Il.: The Univ. of Illinois Press, 1977, pp. 83-86.

entail universal laws. To avoid the difficulty we have to accept certain universal terms which do not entail any universal law but stand for certain observable kinds among which universal laws hold, and theoretical kinds are reduced to these observable kinds. To accept this position is to accept observational-theoretical distinction. But we are not forced to accept this position if the above-mentioned circularity is not vicious. And such circularity may not be vicious if any holistic theory of meaning makes good sense. However, the question is: Does the rejection of observational-theoretical distinction necessarily lead us to accept some form of circularity?

It may be maintained that we end the aforesaid regress not by arriving at certain universal terms denoting purely observable properties, but by arriving at certain universal terms entailing certain laws which have not been formulated in language by using universal terms. This position avoids circularity, but still remains consistent with the rejection of observational-theoretical distinction on the ground that all universal terms entail universal laws. Popper, however, does not adopt this position. Instead he holds that the above-mentioned circularity is inevitable, as we cannot break the circle by establishing a definite order of reduction or introduction.[5] The reason is that by reduction or introduction we cannot arrive at primitive predicates — primitive in the sense that they are non-dispositional or less dispositional than others. All predicates, according to him, are equally dispositional. This position is highly counter-intuitive. It is, for instance, difficult to accept that the term "red" is as much dispositional or theoretical as the term "electric field." In fact,

5. K.R. Popper, *Conjectures and Refutations*, London: Routledge & Kegan Paul, 1972, p. 278.

Popper himself, in a different context, admits that though all terms are theoretical to some degree, some are more theoretical than others.[6]

The more serious difficulty is that Popper's rejection of observational-theoretical distinction clashes with his falsifiability thesis. He holds that a so-called observational term, in addition to directly observable occurrent property or properties, refers to dispositional or causal properties. Causal relation holds between a directly observable property and some other property or properties in the sense that anything having the directly observable property gives rise to the dispositional property or properties under a specific circumstance. Now, if falsifiability be the mark of scientific statements and causal laws are falsifiable, then the causal relation that holds between a directly observable property and a certain dispositional property is also falsifiable. In that case we need a language that is rich enough to describe the causal relation and to accommodate the possible falsification of the causal law. If our language contains no term to refer to only the directly observable property, such a language cannot accommodate the falsifications of certain causal laws.

Consider Popper's own example which he sets forth to explain and criticize the tendency to immunize a theory or law without explaining the new fact which contradicts the theory or the law.[7] The law "Fire produces smoke" is falsifiable, but we cannot say that it is falsifiable if the term "fire" does not refer to anything which does not produce smoke.

Imagine that all our hitherto observed instances of fire

6. K.R. Popper, *Conjectures and Refutations*, p. 119.

7. K.R. Popper, "Autobiography of Karl Popper", in P.A. Schilpp (ed.), *The Philosophy of Karl Popper*, Illinois: Open Court Press, 1974, p. 33.

were associated with smoke, and that we have now confronted an instance of fire not producing smoke and discovered that fire cannot produce smoke in the absence of other conditions like wet fuel. But under the constraint that anything which does not produce smoke cannot be referred to by the term "fire," we cannot even say "Fire cannot produce smoke unless there are other conditions such as wet fuel." Under the assumption that anything which does not produce smoke is not fire, the statement "Fire cannot produce smoke" becomes contradictory. The assertion 'Fire cannot produce smoke unless there are other conditions such as wet fuel" makes sense only if "fire" refers to certain directly observable properties but does not include production of smoke as a part of the meaning or the definition of the term.

It may be true that we normally use a universal term to refer to a cluster of directly observable and dispositional or causal properties. This is perhaps due to our expectation that the causal or dispositional properties of anything having a particular directly observable property will occur under appropriate circumstances. But if our expectation is defeated and if the situation demands, the meaning of the same term is changed and it is used to refer to only the directly observable property. Without this sort of flexibility our linguistic framework cannot take into account the falsifications of causal laws and accommodate new facts.

From the above account it appears that the rejection of the observational-theoretical distinction is incompatible with the falsifiability thesis. Naturally, the rejection of this distinction requires the rejection of the view that causal laws and theories are falsifiable, at least straightforwardly. It is maintained that the experience of a counter-instance falsifies a law or a theory. But Quine holds the view that no particular statement, taken in isolation from its fellows, is directly linked with any particular experience in the sense that it is verified or falsified by any particular experience.[8] According to him, our statement about the external world faces the tribunal of experience not individually but only as a corporate body forming a total system. Any conflict with experience affects the equilibrium of the system as a whole. And in the face of any experience affecting the equilibrium of the system, any statement can be held true if we make necessary adjustments elsewhere in the system. Since it is not significant to talk about verification or falsification of a particular statement taken in isolation, there is no synthetic statement in the sense that its truth or falsity is determined by experience; and consequently there is no analytic statement that is confirmed no matter what.

However, whatever be the reason — pragmatic or non-pragmatic, Quine allows re-evaluation of a particular statement in the face of a recalcitrant experience.[9] But, such re-evaluation, for instance, of the statement "Fire produces smoke" is not possible if it is held that the term "fire" does not refer to anything unless it produces smoke, or that anything which does not produce smoke is not fire. The re-evaluation is possible only if "fire" refers to something directly observable and not its dispositional or causal property.

Even if we make necessary adjustments in our conceptual scheme whenever we face contrary experience, we would not be able to do away with observation terms. How can we accommodate, without re-evaluating the statement "Fire produces smoke," the experience of something that has all the observable properties of fire but not producing smoke. The

8. W.V.O. Quine, "Two Dogmas of Empiricism", in A.P. Martinich (ed.), *The Philosophy of Language*, New York: Oxford University Press, 2001, p. 57.

9. Ibid., p. 58.

only way is to coin a new term, call it "fire$_o$," which refers to all the directly observable properties of fire except the causal property of production of smoke, and then assert that fire$_o$ cannot produce smoke unless there are other conditions like wet fuel.

What, then, would be the status of the statement "Fire produces smoke" in our conceptual scheme? As a scientific statement it is useless. Science is concerned with explanation and prediction. But the statement neither adequately explains the production of smoke nor helps us to predict smoke efficiently.

From the foregoing discussion it may be evident that sometimes the situation demands observational terms regardless of whether we accommodate any contrary experience by making necessary adjustments in our conceptual scheme or by re-evaluating a particular statement. In these situations the understanding of the meaning of a universal term without relating its reference with the dispositional or causal properties becomes necessary and, therefore, the holistic understanding fails.

11

On Inter-Theoretic Incommensurability in Science

Milan Desai

AT the root of the incommensurability syndrome in science, lies the thesis of theory-dependence, viz. that the application of terms in science depends upon scientific laws and theories. One of the seminal figures in developing this semantic position for the language of science, is Karl Popper. Popper, of course, is renowned for his falsificationist methodology. Lesser known are his views on language; partly because these are found scattered in his works, and largely because of Popper's explicit rejection of the Positivists' penchant for meaning/conceptual analysis. But one of the major points this paper seeks to make is that ironically enough, Popper's semantic views, carefully garnered and pieced together, undermine his falsificationist/ empiricist methodology, and lead inexorably to the Kuhnian and Feyerabendian positions of paradigmatic/theoretic incommensurability in science.

The crucial statement of Popper's position is found in the *Logic of Scientific Discovery*. Referring to the view that science is the systematic presentation of our sense-experiences Popper maintains:

> This doctrine founders in my opinion on the problems of induction and of universals. For we can utter no statement

> that does not go beyond what can be known with certainty "on the basis of immediate experience" (This fact may be referred to as the "transcendence inherent in any description"). Every description has universal names (or symbols or ideas); every statement has the character of a theory or a hypothesis. The statement "Here is a glass of water" cannot be verified by any observational experience. The reason is that the universals which appear in it cannot be correlated with any specific sense-experience (an "immediate experience" is only *once* "immediately" given; it is unique). By the word "glass," for example, we denote physical bodies which exhibit a certain law-like behaviour; and the same holds for the word "water." Universals cannot be reduced to classes of experiences they cannot be "constituted."[1]

Popper's argument for the theory-ladenness of universal terms assumes the form of a criticism of the nominalistic theory of meaning which construes non-logical predicates as names. This view interprets universals as "extensionally or enumeratively" defined; but, according to Popper, such a theory of meaning is totally inadequate for the language of science because statements in an extensionalist language can only be analytically true or false. Thus Popper says:

> Nominalism is the doctrine that all non-logical words are names either of a single physical object, or shared by several such objects. This view may be said to interpret the various words extensionally or enumeratively; their "meaning" is given by a list or an enumeration of the things they name.[2]

Such a purely extensionalist language is useless for science

1. K.R. Popper, *The Logic of Scientific Discovery*, London: Hutchinson and Co. Ltd., 1972, p. 94.
2. K.R. Popper, *Conjectures and Refutations*, London: Routledge & Kegan Paul, pp. 262-64, 1969.

because the truth or falsity of its sentences can be decided simply by comparing the defining lists or enumerations, i.e. as soon as the words occurring in it have been given their meaning. Popper concludes that the language of science must make use of genuine universals, i.e. of words, whether defined or undefined, with an indeterminate extension, though perhaps with a reasonably definite intensional meaning. This intensional meaning of all "genuine non-extensional universals" is determined, according to Popper, by theories. This semantic thesis of theory-dependence has disastrous consequences for both verificationism and falsificationism. If no statement in science, in particular not even singular "observational" statements, can be justified by experience, then no hypothesis in science is empirically testable. A fortiori no two competing hypotheses can be tested against experience, i.e. there are no crucial experiments and science is therefore not even negatively under the control of experience.

Popper himself, in later years, concedes that the growth of knowledge takes place only against a theoretical background of accepted beliefs which constitute the "observational theories" with which "observational statements" in science are impregnated. Thus Popper maintains that Kant was right in teaching that any growth of knowledge needs a theoretical framework which must precede the growth.[3] But Kant was wrong, according to Popper, in believing that this conceptual framework could not possibly be transcended in its turn. Again Hegel was right in pointing out that the framework too was subject to growth; and could be transcended. But he was wrong in suggesting that truth is essentially relative to some framework; and that it is not our own active criticism which forces a change in ideas and beliefs;

3. Popper, K.R., *Realism and the Aim of Science*, London: Hutchinson and Co. Ltd., 1983, pp. 153-57.

rather than the criticism being dependent upon an independently evolving framework. The latter view leads, according to Popper, to the "myth of the framework," i.e. to the pessimistic Kantian doctrine that we are hopelessly enslaved by the conceptual framework we (physiologically) inherit. A special form of this philosophy of human bondage is linguistic relativism (*a la* Benjamin Lee Whorf): the thesis that human languages incorporate in their structures beliefs, theories and expectations, from whose ideological fetters we cannot break out by criticism, because any criticism presupposes language. But Popper thinks that although the thesis of theory-ladenness indicates that we cannot do altogether without some form of theoretical framework, yet (relative) freedom can be attained through criticism, both immanent and transcendent, of the most varying frameworks.[4] Yet when we bring to bear upon the issue the foregoing analysis of the semantic function of conceptual frameworks, then it is clear that shifts in framework from one epistemic context to another, do involve semantic variance and incommensurability. This is Feyerabend's thesis.

Feyerabend emphasizes that firstly, the choice of an observational theory is arbitrary.[5] It usually indicates an irrational preference for the "older entrenched" theory and an equally unwarranted bias against the "younger" one. The test of a new theory against such an observational background theory displays at best an inconsistency, which could as well be remedied by jettisoning the observational theory instead of the theory under test. Interpreted as a refutation and

4. Popper, K.R., *Realism and the Aim of Science*, London: Hutchinson and Co Ltd, 1983, pp. 15-30.

5. P.K. Feyeraband, "The Rationality of Science (from "Against Method") in S. Harding (ed.), *Can Theories be Refuted?*, pp. 289-315.

rejection of the new theory, this process renders the context of testing as irrational as the context of discovery. Secondly since the observational theory constitutes the principle of application of terms in the test situation, it defines the meaning of these terms in the context. Two tests of a theory, therefore, against the backdrop of different observational theories are semantically incommensurable. Moreover, in the case of cosmological theories like classical mechanics and special relativity, their tests against a common observational theory (which might even be the physiological theory incorporated in our sensory apparatus) are incommensurable, because the cosmological theories override the observational base, and interpret its concepts in their own terms.

The foregoing analysis makes it clear that the semantic thesis of theory-dependence, viz. that all concepts in science are intensionally defined by accepted, background "observational" laws and theories, leads to the incommensurability syndrome. Does the "new" theory of reference developed by Kripke, Putnam et. al. alter the situation? The new theory is partly a reaction to the semantic tradition of Russell & Frege, which construed names and general terms as attenuated descriptions and considered the reference of these terms to be a function of their Fregean sense. In contrast the modern semanticists argue that names have no intension in the traditional sense; and that at least general terms which designate natural kinds (naturally occurring species) are like names and thus do not have their extensions determined by concepts/intensions. Instead reference is achieved by something like a causal chain rather than by associated descriptions.

Kripke emphasizes that general terms which designate natural kinds, i.e. terms in science, are rigid designators which name the same kind or substance in all possible worlds.[6] Associated properties, therefore, are not criterial, and the statements attributing such properties to individuals in the extension, are not analytic. Furthermore if the terms of an identity statement are rigid designators, the identity statement, if true at all, is necessarily true. But it is neither analytic nor a priori. Thus Kripke distinguishes between the metaphysical notions of necessity/contingency, the epistemic notions of a priori/a posteriori and the linguistic notions of analytic/synthetic. This analysis, when applied to statements of theoretical identity like "water is H_2O" or "gold is the substance with atomic number. . ." implies that (i) the statements if true are necessarily true; (ii) they are not known a priori for they are matters of scientific discovery; and (iii) the statements are not analytic, for the scientific theory might be falsified, and in any case, if true, is true on account of the way the world is, and not on account of any fact about language. *Prima facie* it might seem that the scientific theory associated with a natural kind term determines its extension, if it is the true theory of the kind. But in the absence of this utopian ideal of privileged access to ontological realities, what is the role of "current best theory" Schwartz explains the position in the following manner.[7] He says Kripke and Putnam hold, for examples that water is necessarily H_2O; H_2O is the true nature of water. "Water" rigidly designates H_2O regardless of what superficial properties the H_2O might or might not have. Schwartz says that at this point it is very easy to confuse the new theory with a possible refinement of traditional theories. Such a confusion would ensue

6. S. Kripke, *Naming and Necessity*, Oxford: Basil Blackwell, 1980.

7. S. Schwartz, Introduction to "Naming, Necessity and Natural Kinds", in S. Schwartz (ed.), "Naming, Necessity and Natural Kinds", Ithaca and London: Cornell University Press, 1977, pp.13-41.

if we thought that "Water is H_2O" is analytic. It might be thought that Kripke and Putnam are merely trying to replace ordinary definitions with scientific ones so that instead of defining water as a clear, colourless liquid, we define it as H_2O. This is not the view of the new theory. One would come closer to the position of Kripke and Putnam if one simply said that "water" has no definition at all, at least in the traditional sense; and is a proper name of a specific substance. The reason why the "current best theory" cannot constitute the intension, cannot generate analytic truths is that firstly, it could turn out to be false, and secondly that scientific theories are a matter of scientific discovery and not conventional definition. Hence the current best theory generates, if anything, only an epistemic certainty or a prioricity; the certainty of a well-established empirical theory, not the necessity that ensues from access to metaphysical or ontological realities. Nevertheless, since theories are a matter of scientific discovery, not of definitions, it would appear that scientists are at least approximating to metaphysical, i.e. necessary truths.

The foregoing are the views of Kripke, Putnam et al. Are these views justified? To me, there appears to be a naïve if not desperate clinging to faith in the scientific enterprise as one of empirical discovery, of uncovering the essence of things, the structure of Reality; a faith which is scarcely warranted by the philosophical analysis of science and its language. It is perhaps a deeper appreciation of this contretemps, which has made several philosophers of science develop the thesis of "primitive" classification. "Primitive Classification" in the context of science, means simply that the extension of a natural kind term is not intensionally defined, but only extensionally exemplified. This is Kuhn's thesis of learning by exemplars. Thus Kuhn asserts that the principles of classification are never explicitly articulated; instead in everyday contexts one learns to apply terms based on the implicit recognition of resemblances and dissimilarities between objects which are extensively (indexically) indicated.[8] Similarly, in the context of science, exemplars which are concrete problem solutions as well as direct exposure in the course of laboratory work, teach the student the application of scientific terms. Presumably such application is also based on the implicit recognition of similarities and differences.

Popper points out that one of the main characteristics of similarity is its relativity.[9] Two things which are similar are always similar in certain respects, and therefore may always be similar in different respects. In other words similarity presupposes the adoption of a point of view, i.e. a theoretical standpoint, hence cannot constitute a principle for primitive classification.

At this juncture, we may briefly glance at the views of Quine, at once hard core empiricist and conventionalist. The reason for Quine's strange contretemps is the same as that for Popper's, viz. the lack of an adequate theory of meaning. In the "Two Dogmas of Empiricism" Quine rejects the Aristotelian (and Platonic) notion of meaning as essence.[10] In "Naturalized Epistemology,"[11] he unreservedly embraces the more sophisticated version of the verification theory of meaning

---

8. T.S. Kuhn, "Second Thoughts on Paradigms" in F. Suppe (ed.) *The Structure of Scientific Theories*, Chicago: University of Illinois Press, 1977, pp. 459-82.

9. K.R. Popper, *The Logic of Scientific Discovery*, London; 1972 pp. 420-41.

10. W.O. Quine, "Two Dogmas of Empiricism" in S. Harding (ed.) *Can Theories be Refuted*, 1974, pp. 41-64.

11 Quine's views have been retrieved from J. Kim, "What is Naturalized Epistemology?" in E. Sosa & J. Kim (eds.), 2000, *Epistemology — An Anthology*, U.K.: Blackwell Publishers, pp. 301-13.

whereby it is only the corpus of knowledge as a whole that has empirical significance. This is because, as a conventionalist, Quine believes that the meaning of individual terms and even whole sentences is implicitly defined by the theories in which they occur. Since theories themselves are heavily underdetermined by the evidence, their acceptance must rest largely on consensus and agreement and only partially in the way the world is. Obviously this sets the stage for cultural, linguistic and theoretical relativism and incommensurability.

The foregoing analysis makes it clear that neither intensionalist nor extensionalist nor yet, conventionalistic theories of meaning evade the pitfalls of incommensurability. A fortiori they fail to explicate the central problem of epistemology, viz. the relation between scientific, i.e. empirical knowledge and experience. Towards an attempt at a more satisfactory resolution we must step outside the domain of Western theories of meaning; and glance briefly at the outlines of Bhartṛhari's *sphoṭa* doctrine. It must be made clear, however that it is not to the purpose at hand, to make a detailed or a critical exposition of the theory; but only to indicate tentatively, how any theory of its kind (in the relevant aspects) may help to retrieve (the philosophy of) science from the quagmire in which it is currently enmired, and to show how scientific, i.e. empirical knowledge is possible.

In *Indian Theories of Meaning* Kunjunni Raja says:

> In the *sphoṭa* doctrine, Bhartṛhari gives a penetrating and minute analysis of the speech situation. First, we have the actual sounds of the words uttered, this is the *vaikṛt dhvani*. The *prākṛt-dhvani* is an abstraction from the various *vaikṛt-dhvani*s, or what may be considered as the linguistically normal form devoid of the personal variations which are linguistically irrelevant. The third stage is the *sphoṭa* which

> is the whole utterance considered as an integral unit, as an indivisible language symbol. It is this *sphoṭa* that reveals the meaning which is in the form of an intuition. Strictly both the *sphoṭa* and the meaning are different aspects of the same speech principle.[12]

Kunjunni Raja illustrates the speech situation graphically.

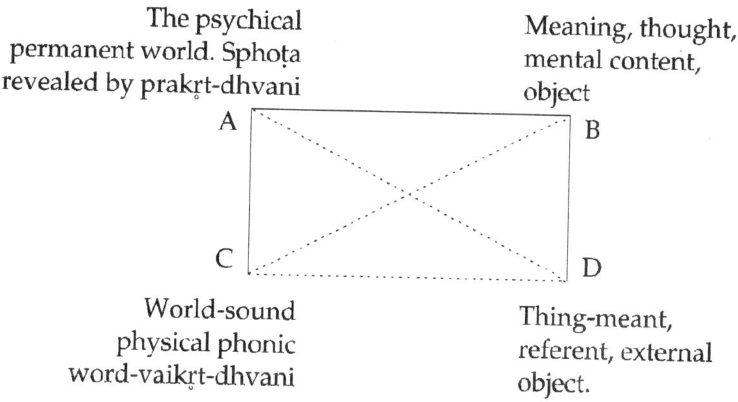

Kunjunni Raja says that there is a direct relation only between A and B in the diagram. Relations between C and D, C and B, and A and D are all indirect and imputed.

The foregoing account of the *sphoṭa* doctrine is sufficient for our purposes. Clearly meaning is distinct from referent/s, i.e. from things or facts. Hence the theory rejects an extensionalist account of meaning. Equally clearly, meaning is not to be identified with any empirical law or theory, for it is revealed by intuition or *pratibhā*, and not by any sensory observation. Hence the theory avoids the pitfalls of theory dependence. But most importantly how does the theory address the problem of the relation between empirical statements and experience? Here suggestive extrapolation and

12. K. Kunjunni Raja, *Indian Theories of Meaning*, Madras: The Adyar Library and Research Centre, 1963, pp. 14-15.

development is indicated, which also invokes the general background of Indian theories of Meaning:

Firstly meaning is clearly distinct from referent/s. Meaningful utterances are related to referents as things or facts meant. It is now suggested that meaningful utterances may be classified as empirical or non-empirical depending on whether the referents are observable or not. Empirical sentences may in turn be classified as true or false. A true utterance will then be defined as just the one whose referent happens also to be the object of experience.

The situation could be graphically illustrated as follows:

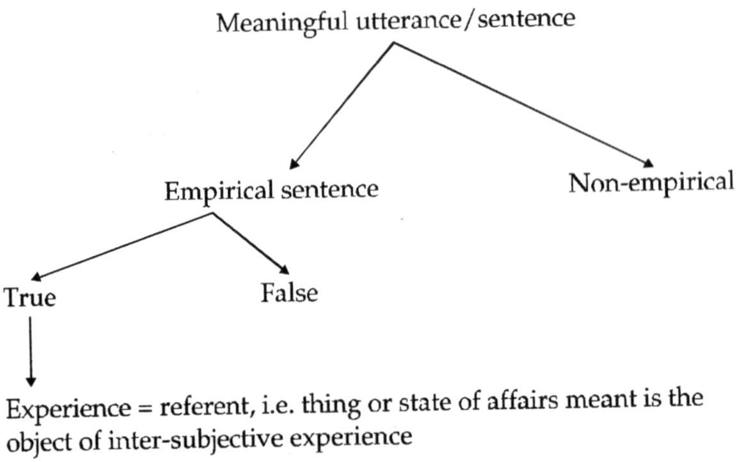

Thus the thing or state of affairs meant by a true sentence would also be the object of inter-subjective experience; whereas the thing or state of affairs meant by a false sentence would not so be. This account suggests that the relation between true sentences and experience is indirect — they are related via the referents, which is also the object of inter-subjective experience. Thus the general relation between language, i.e. meaningful utterances and the world is neither logical nor causal. It is a special linguistic or semantic relation of (meaningful) utterance to thing/fact meant. The world is also related to experience by means of a special relation, viz. that of being the object of inter-subjective experience.

The foregoing is only a very tentative exploratory and suggestive account of how a semantic theory of the *sphoṭa* type which distinguishes sharply between meaning and fact, between meaning and empirical laws and theories, may help to sort out the imbroglio in the philosophy of science in particular, and in epistemology in general.

12

Creativity and the Growth of Scientific Knowledge

Manjari Chakrabarty

In his *Logik der Forschung* Karl Popper made an explicit distinction between the psychological processes which occur when a scientist conceives a new idea or theory and the methods of examining it logically. The act or procedure by which a new theory is generated, Popper argued, does not call for any rational reconstruction and, therefore, falls outside the proper domain of philosophy of science. What held fascination for him instead is the rigorous critical examination and corroboration of "finished" theories, i.e. theories supposed to have already been created. In other words, the thought processes involved in the genesis of a hypothesis were of no relevance to him as he was mainly concerned with the post-discovery method of testing, i.e. post-generational justification. Popper bequeathed the study of discovery or theory generation and consequently the creative aspect of scientific research to psychology and history. Following Popper, philosophers of science, for an entire generation, have left all scholarly, analytical concern with scientific discovery to the psychologists or historians and have taken as their central preoccupation the task of formulating a logic of post-hoc theory evaluation.

The most creative aspect of scientific work has generally been in the development of theories. It is interesting to note that the word "theory" and the word "theatre" have their origin in the Greek word *theoria*, which means "to view" or "to make a spectacle."[1] This implies that primarily a theory is a way of looking at the world through the mind. It is thus widely regarded that the conceiving of a scientific theory having a broad and deep significance epitomizes the highest plateau of creativity. Now one might naturally expect that philosophers of science (concerned centrally with the structure of science and scientific thought) will have interesting and important suggestions about the nature of creativity. Quite the contrary there has been a hesitation on their part to explore the thought processes involved in the conceiving of a scientific theory.

This discovery-justification dichotomy has received little critical attention until the stimulating challenge by N.R. Hanson's *Patterns of Discovery* (1958). In the past two-three decades some philosophers of science have become sensitive to the fact that to exclude the processes of discovery and consequently the creative dimension of scientific research from their domain of professional interest amounts to neglect of highly relevant phases of scientific investigation. In this paper I try to show how a little more attention to the processes of theory generation might have helped Popper avoid a number of critical attacks in relation to his views on the growth of scientific knowledge. This implies that I intend to overthrow the Popperian claim that there is nothing in the construction of scientific hypotheses to concern the philosophers of science.

Popper, as we all know, believed that the great scientists

---

1. D. Bohm, *On Creativity*, London: Routledge, 1998, p. 43.

are often guided by "creative intuition" or inspiration, rather than by rational thought, in arriving at a new idea or theory. As every discovery contains an "ineliminable psychological element," the discovery process, i.e. the process of generation resists logical analysis and thus is of no significance to him. This Popperian view of discovery as an unanalysable psychological affair permeated by intuitions or inspirations, seems to be deeply influenced by what has been dubbed as the "classical view"[2] of creativity. This classical view according to which the act of creation is mysterious, ineffable, places great emphasis on sudden moments of inspiration or flashes of insight and it gains support from the introspective self-reports by acknowledged scientists like Poincare, Helmholtz, Kekule and others. These self-reports are well-documented accounts of those "eureka" moments or what Popper calls moments of "creative intuition" which accompany the dawning of deep and significant scientific discoveries.

This classical view, which dominated literature on creativity for long, has been severely criticized by contemporary creativity researchers. For instance, Boden has showed that from the Popperian thesis (that scientific discovery cannot be predicted) it does not necessarily follow that nothing can be said, in general terms, about the mental processes by which new ideas arise.[3] Perkins, on the basis of his empirical studies, has criticized this traditional view that new ideas in science are always generated irrationally.[4] Schaffer has

---

2. See S. Dasgupta, *Technology and Creativity*, Oxford: Oxford University Press, 1996, p. 88.

3. M. Boden (ed.), *Dimensions of Creativity*, Cambridge, MA: The MIT Press, 1994, p. 3.

4. D.N. Perkins, "Creativity: Beyond the Darwinian Paradigm", in M. Boden (ed.), *Dimensions of Creativity*, Cambridge, MA: The MIT Press, 1994, pp. 119-42.

commented that Popper's distinction between discovery and justification is "over simple."[5] There has also emerged in recent times a new breed of investigators, who armed with analytical tools derived from cognitive psychology and computer science, have begun to probe into the cognitive processes of scientific discovery. As a result of their investigation into particular historical cases, they have come to suspect the validity of the very assumption that there is nothing in the context of discovery to concern the philosophers of science. My aim in this paper is not to get involved into the debates on whether or not there can be a logic or specifiable rational procedure for great scientific discoveries. In what follows I shall attempt to show how studying the mere testing and corroboration of "finished" theories and ignoring the processes of theory generation have led Popper face a number of difficulties in relation to his views on the growth of scientific knowledge.

First of all, let us concentrate on the term "discovery." Popper has not been very clear in what sense he used this particular term. While reading Popper one easily gets the impression that rather than treating discovery as a complex development extended in both space and time he has construed discovery as concerned with the "eureka" moment, i.e. the moment when a new idea or hypothesis first dawns. However, closer examination shows that Popper was at least been dimly aware of a distinction within the discovery-context between the processes involved in the stimulation and release of an inspiration and the processes of critical examination involving subsequent tests by the scientist himself ". . . whereby the inspiration may be discovered to be a discovery."[6] Though

5. S. Schaffer, "Making up Discovery", in M. Boden (ed.), *Dimensions of Creativity*, Cambridge, MA: The MIT Press, 1994, pp. 13-51.

6. K.R. Popper, *The Logic of Scientific Discovery*, London: Routledge, 1992, p. 8.

Popper had an inkling of the fact that a kind of evaluation or assessment, at least on the part of the scientist himself, is involved in the context of discovery, he did not develop the matter any further.

This traditional, rigid distinction between the discovery context and the normative, reconstructive justification context proposed by Popper has lost much of its philosophical interest by being formulated in many different ways by later philosophers of science.[7] A group of researchers prefer to view science in the large as a complex three-stage process, viz. generation, pursuit, and acceptance, and they use to distinguish one sense of "discovery" corresponding to each phase of research. At the generation stage "discovery" means the initial formulation of an idea prior to the question of explicit assessment. In a somewhat stronger sense the term "discovery" applies to the production of an idea which is pursuit-worthy, i.e. worthy of some expenditure of time and effort by the relevant scientific community. Unlike the former "discovery" in this sense involves active consideration by researchers but not necessarily acceptance. Finally we have "discovery" as a scientific achievement accepted and certified by the scientific community.

This pursuit phase (which is sometimes referred to as "preliminary evaluation") applies to the situation in which a theory or theory sketch has been generated but it has not yet been developed to the point of final testing. At this stage the theory needs to be articulated, modified and clarified. Clearly this phase is to be distinguished from discovery in the sense of initial thinking or generation, because the basic ideas are already present. Neither does it fall squarely within the

---

7. See, for example, T. Nickles (ed.), *Scientific Discovery, Logic, and Rationality*, Dordrecht: D. Reidel Publishing Co., 1980, pp. 1-22.

purview of final acceptance or justification, as the theory is still in the developmental stage and is not yet ready for rigorous testing. As it falls between the generation phase and the acceptance phase science can be viewed basically as a three-stage process. This separation of theory generation from pursuit or preliminary evaluation must not lead anyone to assume that generation is devoid of all evaluation. The reason I draw attention to this three-stage view of science is that the traditional discovery-justification dichotomy apparently neglected this pursuit phase of scientific research; and this might have led the critics to comment that Popper portrays "discovery" as a momentary affair, as an "*ahā*" experience. It is perhaps not too difficult to uncover why Popper did not pay sufficient attention to the pursuit phase of scientific research.

This pursuit phase, in which ideas developed within a framework are fleshed out, articulated and extended, is comparable to what Kuhn calls "normal science." Popper never liked Kuhn's idea of "normal" periods of intensive, convergent research, dominated by one ruling theory or paradigm as being characteristic of science. However imperfectly Kuhn might have illustrated the concept of normal science, it seems, his merit lies in pointing out a very crucial and then neglected aspect of scientific research. There is evidence that even the scientists themselves have acknowledged the importance of this pursuit phase of scientific investigation. Planck, for instance, has stated that most of the ideas that play a part in science are not perfect when they first take shape; rather they need to be articulated and their perfecting is a gradual process.[8] This distillation or refinement of existing ideas into more

8. M. Planck, *The Philosophy of Physics*, USA: Norton Library, 1963, p. 115.

embraceable forms is a continuous and crucial part of the scientific enterprise. In Bohm's view new theories are like growing plants that need to be nurtured and cultured for a time before falsifiable inferences can be drawn from them.[9]

Popper, impressed by the bold ideas and the highly critical activities of the great scientists, portrayed science as a process of trial and error, of proposing bold, new conjectures and then subjecting these conjectures to severe critical discussions and empirical tests. Critics find this picture of science presented by Popper much too simple to be true. This simple view of science might make us think that Popper did not pay adequate attention to the need of intensive convergent work of articulating and modifying those conjectures before they can be put to rigorous tests. Although it is not quite right to believe that Popper was totally unaware of the necessity and significance of this part of scientific research as some of his comments do suggest the contrary, we have reason to argue that he has somewhat devalued this pursuit phase of scientific investigation. His lack of scholarly interest in this important aspect of scientific development seems to be a necessary corollary of his notion of scientific discovery as an intuitive, psychological affair.

Popper devoted much attention to criticize the "narrow inductivist conception" of scientific enquiry which entails that scientists are in possession of a method that enables them to infer scientific theories mechanically from an exhaustive collection of facts gathered without any theoretical preconceptions. Against this he held that scientific theories are "free creations of the human mind," they are pure guesses or genuine conjectures. In Popper's view, the creation of scientific theories such as those associated with the names of

9. D. Bohm and D. Peat, *Science, Order, and Creativity*, London: Routledge, 1989, p. 59.

Newton and Einstein involves an essential element of originality which could never be produced by following an algorithm. Thus Popper strongly denied that anything like a logic of discovery is possible.

Let us look at the matter a little closely. Whenever Popper spoke about discoveries he usually had in mind the generation of such theories as universal gravitation, relativity and so on which Penrose refers to as "superb."[10] Now, all these theories involved radical conceptual changes into their respective domains, they eventually transformed the whole framework of the traditional mode of thought. Carried away by the revolutionary nature of such great theories, which, as a matter of fact, are quite rare, Popper might have proceeded to reject the possibility of any logic of such revolutionary discoveries. The argument of Popper, however, is a general one — ". . . there is no such thing as a logical method of having new ideas. . . ."[11]

Critics, for instance, Kuhn rightly observes that whenever Popper talks or writes about scientific discovery he tends to dwell only on the great names and the great events. But there exist significant differences between the processes of revolutionary discovery and that of small-scale or normal discoveries, which, Kuhn believes, Popper did not take account of. In his paper, "Does Scientific Discovery have a Logic?," [12] Simon also notices that those who have argued against the possibility of a logic of discovery have drawn no clear distinction between normal and revolutionary discoveries and

10. R. Penrose, *The Emperor's New Mind,* Oxford: Oxford University Press, 1989, p. 152.

11. K.R. Popper, op. cit., p. 8.

12. Published later in H.A. Simon, *Models of Discovery,* Dordrecht: D. Reidel Publishing Co., 1977, pp. 326-37.

then shows with the help of examples how a logic of small-scale or normal discoveries can be constructed.

It is a fact that we do not yet have, and probably will never have, anything like algorithmic rules for generating such novel, revolutionary scientific theories. Thus there seems no strong reason to completely reject the Popperian view. However it would be a great mistake on our part to ignore the enormous progress made in recent years in what Nickles calls the "logics" of discovery.[13] More importantly, it can be argued against Popper that in order to deny the possibility of a logic of scientific discovery one must not necessarily hold a "mystical" view of theory generation. Popper seems to have thought that there are only two alternatives: the narrow inductivist conception of scientific enquiry, and the view of scientific discovery as an intuitive, unintelligible process. Setting aside the possibility of an algorithmic procedure for theory generation Popper went on to make the processes of discovery dependent largely on intuition and inspiration which cannot be made intelligible to reason. In the light of the writings of later philosophers of science and creativity researchers the argument of Popper seeking to prove the impossibility of a logic of scientific discovery is likely to be received with skepticism.

There is need to invoke history here. Several philosophers of science and creativity scholars have strongly suggested on the basis of their empirical studies that reasoning and logic may play a key and ubiquitous role in the discovery process. For example, Hanson has pointed out in one of his articles that the proposal of a hypothesis can be a reasonable affair and is not as dependent on intuition as philosophers suppose.[14]

13. T. Nickles, op. cit., p. 26.

14. N.R. Hanson,"Is There a Logic of Scientific Discovery?", in B.A. Brody (ed.), *Readings in The Philosophy of Science,* New Jersey: Prentice-Hall, 1970, p. 628.

However, these reasons may differ in type from those which lead one to accept a hypothesis once suggested. Lakatos' idea of a research programme, characterized in part by its "positive heuristic" can be seen as an attempt to highlight this aspect of science more precisely and in more detail. Laudan has also accepted that research traditions can play a vital heuristic role in the construction of specific theories. The introduction of a new hypothesis, Shapere has claimed, is itself a rational matter and characteristic reasoning patterns can be found for justifying their introduction, exploring and developing them.[15] Succinctly stated, the point is that there is room for rational heuristics in physical science. My intention is neither to underestimate the powers of intuition and imagination, nor to suggest any such idea that scientific theories can be produced mechanically. I simply attempt to point out that a more sophisticated account of theory generation can illuminate such topics as theory change and the advancement of scientific knowledge, which Popper always found interesting and worth investigating.

It is important to note at this point that in his *Die Beiden Grundprobleme der Erkenntnistheorie* which was written before *Logik der Forschung* but published in 1979, Popper espoused an explicitly "Darwinian" account of theory production. In that early work he pointed out that if new theories turn out to explain certain facts, this is not the result of rational efforts on the part of the scientists, rather it is an effect of the superabundance of available theories. New hypotheses emerge like spontaneous mutations; most of these will be experimentally falsified and rejected; but since there has been an overproduction of hypotheses, some of them will survive. In his 1973 Herbert Spencer lecture entitled "The Rationality of Scientific Revolutions" he again presented a "Darwinian"

15. F. Suppe (ed.), *The Stricture of Scientific Theories,* Chicago: University of Illinois Press, 1977, p. 697.

account of theory production which sees the trials in the trial-and-error method of science as "random" or "blind," that is, not deliberately structured to meet the epistemic environment into which they are to be launched. Just as mutations in Darwinian Theory are not environment directed, nor pre-designed to solve some range of ecological problems, so scientific theories are not pre-designed to meet existing epistemic problems: they are generated in a way that is blind to existing epistemic needs, although, once generated, they are subjected to rigorous selective pressure.

This Darwinian account seems directly at odds with the "rational heuristic view" outlined above. For instance, Perkins strongly criticizes those who see human creativity as a Darwinian process of random generation and selective retention.[16] Contemporary philosophers of science too vehemently argue that the articulation of promising theories is not a question of throwing out possible conjectures "at random" and then subjecting them to rigorous selection pressure. History and a little reflection suggest that scientists do not simply throw bold guesses. In an important way sets of more or less rigorous rules, routines or heuristics for discovery already exist. Worrall's detailed analysis of how Fresnel might have arrived at his new, transverse theory illustrates that there is nothing in this process that even remotely resembles the producing of a bold conjecture following the refutation of an earlier theory. Rather it can be reconstructed as a systematic and logical argument based on background knowledge.[17] In simple words, it is difficult for present-day thinkers to believe that theorizing has no element of goal-directedness.

---

16. D.N. Perkins, op. cit., pp. 119-41.

17. See J. Worrall, "Revolution in Permanence: Popper on Theory-Change in Science", in A. O'Hear (ed.) *Karl Popper: Philosophy and Problems*, Cambridge: Cambridge University Press, pp. 75-102.

To put forward the thesis that theory construction in science is in some sense a rational process is neither to suggest that the process is in any sense deterministic in the way algorithm is, nor to imply that it is a matter of mere (algorithmic) calculation, with no need for imagination and intuition. The researchers who are trying to provide us with a plausible rational reconstruction of the processes of theory generation are perfectly aware of the fact that no theory of creativity can be predictive. In other words, they do not ignore the contingent nature of such an act. Thus it does not seem impossible to reconcile the views of these contemporary researchers with that of Popper's. It is rather his attitude towards the processes of theory formulation, which being psychological and involving non-rational elements appear unanalysable, that discourages him from exploring them. In consequence Popper had to face a number of difficulties which could have been avoided.

Before proceeding any further let me explain first what do we exactly mean when we apply the word "creative" in the realm of science. Generally we think that creativity involves the production of an original, useful product. Thus it has often been defined as the process of bringing into being something novel and useful. At this point I would like to ponder into Dasgupta's precise characterization of creativity.[18] Boden has introduced a distinction between two senses of creativity, namely, "psychological" or P-creativity and "historical" or H-creativity, which relates to the conception of fundamental novelty. Dasgupta has proposed a further refinement of Boden's conception of psychological and historical creativity, and distinguishes between "mere novelty" and "true originality" in both the personal-psychological and communal-

18. See S. Dasgupta, *Technology and Creativity*, Oxford: Oxford University Press, 1996, pp. 53-65.

historical contexts. Novelty and originality, which are frequently considered to be characteristic of the creative solution, are not to be judged in any absolute sense. They are functions of and relative to specific "bodies of knowledge" and the agent or community under consideration. This term "knowledge body" has been used by Dasgupta to mean an integrated and organized collection of facts, theories, rules, exemplars, models, beliefs, metaphysical commitments, and values pertaining to a particular subject matter or domain. Such knowledge bodies, their constituent facts, theories, values, standards, and so on, change over time. What Dasgupta has referred to as "knowledge body" in the natural sciences is comparable to what Kuhn termed "paradigm," or what Lakatos called "research-programmes" or what Laudan called "research tradition."

A knowledge body associated with a particular agent and pertaining to a particular domain is referred to by him as the agent's "personal knowledge body" about the domain. On the other hand, a knowledge body associated with and shared among a community of agents and relating to the same domain is called the community's public knowledge body about the domain. Now any thought product, say X, will be said to be "psychologically novel" or PN-creative for the agent if the agent's personal knowledge body contains no other thought product that is identical to X. If the agent believes that X significantly contributes to the relevant community's public knowledge body associated with the domain, then X will be "psychologically original" or PO-creative for the agent. If in addition the relevant community firmly maintains that there exists no other thought product in its public knowledge body that is identical to X, then X will be held to be "historically novel" or HN-creative for that community. Moreover if the community agrees that X is important in advancing the "state of the art" of its public knowledge body then we shall say that X is "historically original" or HO-creative. A thought product that is considered by the relevant community to be "historically original" represents the highest type of creativity.

Now let us see what is meant by "contributing significantly to the community's public knowledge body" or by "advancing the state of the art of that domain." It is widely agreed among scientists, historians and philosophers of science that a discovery or invention that leads (immediately or eventually) to a major change in the conceptual structure of knowledge in that science or that opens up a new field of study or a new class of research problems contributes significantly to the relevant community's public knowledge body. Such a discovery or invention that appears to be highly original and revolutionary in nature, Dasgupta has convincingly argued, must be regarded as HO-creative.[19] Two crucial points are to be noted here. First, a scientific discovery, an example of what is called HO-creativity, is identified as such not intrinsically, in terms only of the properties of the thought product itself, but largely extrinsically, in terms of what relevant people think about it. This implies that one cannot study creativity by isolating individuals and their works from the social and historical milieu in which their actions are carried out. The second point is more important since what emerges to be a hallmark of the highest type of creativity constitutes the very essence of Popper's theory of the growth of knowledge.

Popper, inspired by the basic tenets of organic evolution through natural selection propounded by Darwin in his *Origin of Species* (1859), subscribed to the evolutionary approach regarding the problem of knowledge. He offered a schematic representation of the growth of scientific knowledge through

19. S. Dasgupta, op. cit., pp. 58-65.

error elimination by way of criticism, which is as follows: P1 — TT — EE — P2. Here "P1" refers to the initial problem situation challenging the scientists. Once the scientists are faced with a problem, they may begin to work on it by attempts of two kinds: they may proceed by first attempting to guess a solution to their problem; and they may then attempt to criticize their guess or conjecture. Thus "TT" stands for the tentative theory proposed as a solution to the problem encountered, whereas "EE" refers to the process of error elimination by way of critical discussions and experimental tests. A guess or conjecture may withstand these criticisms for some time. But as a rule, the scientists soon find that their conjectures can be refuted, or that they do not solve their problem, or solve it in part, and furthermore that even the best solutions — those able to resist the most severe criticisms — routinely give rise to new difficulties, new problems. Thus "P2" refers to the new, revised problem situation that emerges from this trial and error process.

In the light of this schema science appears to be a growing system of problems, as scientific knowledge grows from old problems to new problems, problems of ever-increasing depth and fertility, through error elimination by way of systematic criticism. Implicit in it is the idea that scientific knowledge is a phenomenon that grows not only in content but also in its promise in opening up new avenues for our better understanding of the world. This Popperian notion of the growth of knowledge naturally reminds us of Dasgupta's characterization of HO-creativity. Scientific discoveries are widely regarded as HO-creative when they spawn new fields of enquiry, when they uncover new class of research problems.

In addition to what has been just said Popper always emphasized the revolutionary nature of scientific progress. He was fully convinced that science progresses not by accretion but instead by the dynamic process by which an older theory is overthrown and replaced by a new one. Progress in science, therefore, is non-cumulative or revolutionary in the sense that the new theory (logically) contradicts its predecessor and overthrows or supersedes it. In order that the successor theory should represent a triumph over its predecessor it must, according to Popper, have new and testable consequences which were unobtainable from its predecessor; and it must lead to prediction of phenomena which have not so far been observed. In simple words, the successor theory should contradict its predecessor in several points, and what is more important, in contradicting its predecessor it must be successful.

To cite Popper's favourite example, Newton's theory of mechanics and universal gravitation entails that time is absolute, so that two events simultaneous in one frame of reference are simultaneous in all; that every body acts at a distance on every other body; that the inertial mass of a given body is a (velocity-independent) constant. Einstein's theory of relativity contradicts each of these entailments. According to it any two spatially separated events simultaneously relative to one frame of reference are not simultaneously relative to another frame moving with respect to the first; there is no action-at-a-distance; and the inertial mass of a body increases with its velocity. In sum, Einstein's theory has radically transformed the general tacit infrastructure of ideas that underlied contemporary science. Once again the close link between creativity and the growth of knowledge in the domain of science can be noted at this point. A thought product that leads to such a radical change in the conceptual structure of knowledge in that science, as discussed above, is considered to be "historically original" representing the highest type of creativity.

Progress in science, although revolutionary, is, Popper added, in a certain sense always conservative. (By stating this thesis what Popper actually wanted to reject is the Kuhnian suggestion that successive theories are incommensurable, which implies a break within the evolution of science.) Let us take a closer look at the "principle of correspondence" which has been offered by Popper as an *ex post* appraisal criterion. This correspondence principle is the requirement that, in cases of theory change, the new theory, if it is to gain scientific acceptance, must always share the empirical successes of the old theory. A new theory generally achieves this feat by "going over" to the old in the empirical domain in which the old was successful. Now, critics feel, that Popper clearly admitted to cumulative progress at the empirical level (as the successor theory preserves the empirical success of its predecessor) and at the instrumental level as well (because the successor theory allows us to deal with nature at least as well as the old theory did). Although Popper never used the term "cumulative" while discussing the conservative aspect of revolutionary progress, critics have not hesitated to interpret his thesis as that of progress by accumulation of content.[20] What might have led the critics to this view is Popper's unclear response when challenged to explain this conservative aspect of revolutionary growth.

When asked, how Newton's theory "contains" both Kepler's laws and Galileo's law of falling bodies while the former logically contradicts the latter or similarly how Einstein's theory "contains" Newton's when it logically contradicts Newton's system, Popper replied that this is a very "strange" way of containing.[21] The point is it strikes

---

20. See for example, L. Laudon, *Progress and its Problems*, Los Angeles: University of California Press, 1977.

21. P.A. Schilpp (ed.), *The Philosophy of Karl Popper*, La Salle: Open Court Publishing Co., 1974, p. 1151.

Popper as "strange" because he believed in the "flash of insight" model of discovery. If scientific theories are mere bold guesses produced by intuitions or inspirations and if reason has at all no role to play in the processes of theory construction then it really becomes difficult to explain how, in cases of theory change, the successor theory preserves or shares the empirical successes of the old theory. What is required, it seems, is a more sophisticated account of discovery, which can illuminate such problems as theory change and the growth of knowledge.

We have already seen that contemporary philosophers of science are increasingly coming to realize that there are patterns of reasoning in the construction of scientific hypothesis. Examples from the history of science too lend support to the thesis that reasoning is not antithetical to creativity and may play a key role in the processes of theory generation. The validity of the very idea of sudden insights or intuitions as being the *sine qua non* of the creative act has long been rejected by creativity researchers. Thus we may be allowed to say at this point that the scientists arrive at their hypothesis in a way that, though it may no doubt involve intuition and imagination, can nonetheless be reconstructed as a systematic argument based on previous successes in science. This, in turn, leads us to take a closer look at the influence of "background knowledge" on the generation of new scientific hypothesis.

Fortunately in recent years a few studies have been conducted which demonstrate how specific acts of creation in the domain of science can be explained in terms of a plausible cognitive process. These include, among others, Kulkarni and Simon's development of a computer programme that reconstructs how bio-chemist Hans Krebs discovered the orinsthine cycle for urea synthesis,[22] and Paul Thagard's computational models that explores how theories such as the wave theory of light might have come about.[23] Implicit in these kinds of approaches is the claim that acts of theory formulation, even of the most creative kind, are knowledge-intensive. They entail accessing to and retrieval from a broad and rich spectrum of knowledge ranging from tokens highly specific to the domain to the most abstract, even philosophical propositions. What immediately comes to my mind is one of the brilliant articles of Gerald Holton. Holton in that article shows how the "cultural milieu" in which Einstein found himself has deeply influenced and conditioned his scientific works.[24]

The creative agent, as noted above, draws upon a variety of knowledge tokens during the process of theory generation. One particular kind of knowledge that pervades science is that of the past. In order to explain how some particular scientific idea is born in an agent's mind, it may be necessary to invoke the past. Thus if we examine the history of a particular scientific hypothesis we may discover that it exhibits a phenomenon that can only be described as evolutionary. This has also been admitted by the scientists themselves. Einstein emphasized over and over again that his work was firmly embedded in the tradition of physics and had to be considered an evolution of it. In a letter to Carl Habicht he referred to his work on relativity as an evolutionary act, a

22. D. Kulkarni and H. Simon,"The Processes of Scientific Discovery: The Strategy of Experimentation", *Cognitive Science*, 12 (1988), 139-76.

23. P. Thagard, *Computational Philosophy of Science*, Cambridge, MA: MIT Press, 1988.

24. G. Holton, "Einstein and the Cultural Roots of Modern Science", *Daedalus*, Winter (1998), pp. 1-44.

"modification of the teachings of space and time."[25] However, this view that a novel, original scientific theory has an evolutionary past does not diminish in any way its creator's ability. No sort of calculation from known grounds will suffice for creative production. The problem of the relationship between creativity and knowledge crops up at this point.

Since creative thinking, by definition goes beyond knowledge there appears to be a tension between knowledge and creativity. True creative thinking, according to many, produces results of such novelty that they are totally unrelated to what came before. This theory, which assumes the relationship between knowledge and creativity as one of tension, has a long history in psychology. Recently, a number of researchers involved in re-examining this relationship have stated the opposite of this "tension view," that implies, knowledge is positively related to creativity.[26] Now the main task is to see how past knowledge is extended to new situations and how creative thinking builds on past knowledge. I would like to conclude this paper by drawing attention to some possible suggestions regarding how the scientists utilize the positive contributions that their forerunners have so laboriously made and yet are able to break away from past knowledge, to transcend the tradition.

Drawing on past knowledge does not mean mechanically reproducing what existed in the past. The relationship of a creative scientist to the results of the work of earlier scientists is worth considering here. Such a scientist, to say the minimum, has to perceive and understand the world in a fresh way, has to explore what is unknown. He cannot go on with mere

25. G. Holton, op. cit., p. 22.

26. R.J. Weisberg, "Creativity and Knowledge: A Challenge to Theories", in R.J. Sternberg (ed.), *Handbook of Creativity*, Cambridge: Cambridge University Press, 1999, pp. 226-50.

variations on old themes, leading to modifications, extensions, or other developments within the framework of what has already been known. What is called for is that he "learns" from what his predecessor did in the sense that he not only understands the work of the earlier scientists but, as Bohm explains, "perceives" the difference between past knowledge or ideas and those that are now developing in his own mind as he works on the subject. In the domain of science this perception of relevant differences is the basic step. What is meant by "perception" here is clearly not the routine and mechanical kind but what Bohm refers to as "a genuinely creative perception." In a genuinely creative act of perception one first becomes aware of a new set of fundamentally significant differences between the actual fact and one's preconceived notions and then one begins to note a new set of similarities which cannot be derived from previous knowledge.[27] Such sensitivity to difference and similarity enables the scientist to come up with entirely new conceptual structures, with broad application in many fields of thought and experience.

If one reflects on the history of science one may see that creative scientists like Newton and Einstein have perceived new basic differences and have eventually initiated new order in physics. The ancient Greeks assumed that the earthly and heavenly matters are of different natures. The complicated movements of earthly matter revealed its imperfect nature, whereas the heavenly matter moved in a circle, the most perfect of geometrical figures, which expressed its perfect nature. But when experimental evidence did not disclose that heavenly bodies do, in fact, move in perfect circles the scientists started suspecting that the distinction between heavenly and earthly

27. D. Bohm, op. cit., pp. 1-26.

matter was not a fundamental one. It was Newton who noticed that a much more relevant set of differences exist in the successive states of movement of each particle of matter. He found that, generally speaking, the differences were in the positions and velocities of material bodies at successive moments of time. In empty space the distances covered by such bodies in similar intervals of time were thought to be similar in magnitude and direction, thus leading to motion at straight line at constant speed. In the presence of matter these successive distances and their directions were thought to be different.

A part of Newton's ideas was that space and time differences were universally similar, in such a way that different observers would all agree on what was the same interval of time and the same distance in space. Much later Einstein was able to perceive that the facts available to him could be clearly understood if we suppose that observers going at different speeds are actually attributing the property of simultaneity and of being at the same distance to different sets of events. Einstein's basic step went far beyond mere reordering of existing concepts. Rather he was able to detect a new set of relevant differences, from which there arose a new relationship of similarity (he saw that observers having similar differences in velocity would have similar differences in their ways of choosing the sets of events to which the properties of simultaneity and of being at the same distance were attributed), and thus a new order of space and time.

These examples suggest the contrary of the widespread "tension view" namely, that creative products are not comprehensible in terms of what was known before. More importantly, they might also help Popper in tiding over the difficulty of explaining the conservative aspect of revolutionary discoveries. From a closer analysis of how a scientist conceives a new theory a great deal of illumination of the scientific enterprise can be attained. Had Popper spent a little more effort in studying the processes of theory generation, he could have avoided a number of criticisms.

PART III

Cognition and Mind

13

How did the Faculty of Language Evolve? A Debate between Chomsky and Pinker

Amita Chatterjee

Some simple things shaped for generation after generation
Until it lives in our hands and in our eyes, and it's ours.

— Amitav Ghosh, *The Hungry Tide*

CHOMSKY, along with his two colleagues and collaborators, Hauser and Fitch wrote a controversial review article entitled "The Faculty of Language, What is It, Who has It and How did It Evolve,"[1] which led to the publication of a series of rejoinders and defences on the nature of the faculty of language. In this paper, I shall present and critically evaluate the main arguments of Hauser, Chomsky and Fitch (HCF, henceforth), see whether or in what respects the latest position of Chomsky has deviated from his older theories on the faculty of language (FL) and why I have reservations against the HCF hypotheses.

Controversies abound regarding the nature, function and the mode of evolution of FL. HCF propose conciliation between

1. M.D. Hauser, N. Chomsky, W.T. Fitch, "The Faculty of Language: What Is It, Who Has It, and How Did It Evolve?", *Science*, 298, no. 5598, November 2002.

the theories of FL and the amassed empirical evidences from the *biolinguistic perspective*. The concern is undoubtedly interdisciplinary and requires collaboration from evolutionary biology, anthropology, psychology and neuroscience. While searching for a common ground of collaboration, HCF noted that 'language' means different things in different disciplines. "For many linguists, 'language' delineates an abstract core of computational operations, central to language and probably unique to humans. For many biologists and psychologists, 'language' has much more general and varied meanings, roughly captured by 'the communication system used by human beings.'"[2] HCF find it difficult to locate the point(s) of intersection of these two conceptions of language and hence they simply resort to a union operation — ultimately leading to the fractionation of the faculty of language. In the process of establishing what is special to this faculty and which aspect of it is special to humans, HCF offer three hypotheses or conjectures for empirical verification/falsification.

1. *The dual aspect hypothesis:* It upholds a bifurcation of FL into FLN (narrow) and FLB (broad). The former comprises the computational mechanisms for recursion, while the latter consists of the sensory motor (phonetics/phonology) and the conceptual-intentional system (semantics/pragmatics) and other aspects of language.
2. *The recursion only hypothesis:* FLN and its interfaces with FLB are recursive. This recursive fragment comprising

---

2. R. Jackendoff and S. Pinker, "The Nature of the Language Faculty and its Implication for Evolution of Language (Reply to Fitch, Hauser and Chomsky)", *Cognition*, 97, 2005, p. 180. See also R. Jackendoff, *Semantic Structures*, Cambridge, MA: MIT Press, 1990. R. Jackendoff, *Languages of the Mind*, Cambridge, MA: MIT Press, 1992.

the narrow syntax or the computational core is uniquely human. While traits and functions present in FLB are shared in common with non-human animals, FLN is absent in non-human species.

3. *The exaptation hypothesis:* FLN did not evolve by adaptation for the purpose of enhanced communication but most probably have been exapted away from previous adaptive functions, viz. spatial or numerical reasoning, Machiavellian social scheming and multi-purpose tool-making.

HCF recommend the use of comparative method to trace the evolution of FL. The comparative method was used by Darwin to draw detailed inference about our extinct ancestors from empirical data of their living descendents. The rationale for employing the comparative method is threefold: (a) linguistic behaviour does not fossilize; so fossils, skull shapes, etc. do not tell us much about the evolution of FL; (b) many aspects of neural and developmental functions are conserved across different species, so we should try to detect analogous (i.e. inherited by adaptation to a common function, e.g. similarity in structure between octopus's eyes and human eyes) and homologous (i.e. inherited from a common ancestor) aspects; and (c) to establish FLN as uniquely human, one needs to show that no other animal possesses this trait. What other method can be used for this purpose than the comparative method?

This article surprised both friends and foes of Chomsky because it appears that Chomsky has recanted many of his earlier views here. In *Reflections on Language*,[3] Chomsky claimed that language is a complex ability for which the human brain,

3. N. Chomsky, *Reflections on Language*, New York: Pantheon Press, 1975.

and only the human brain, is specialized. But here he speaks of FLB which non-humans also share. In *Rules and Representations*,[4] Chomsky distinguished "language faculty" from other cognitive abilities. This view was strengthened in his modularity thesis which emphasized that the functioning of the language faculty is guided by domain-specific special principles. Here, on the other hand, Chomsky maintains that the evolution of narrow language faculty from the ability of navigation, etc. may have been possible because what was domain-specific in non-humans has most probably turned out to be domain-general in the human context. This view is startling because Chomsky[5, 6, 7] has been known to maintain that (a) current findings of biology and brain sciences do not provide any support for well-established conclusions about language and (b) computational system of language is biologically isolated since biological systems usually are bad solutions to some design problems. Nevertheless he has joined hands with bio-linguistic approach. However, he still adheres to his favourite LAD (Language Acquisition Device) thesis and tells the evolutionary story in consonance with his Minimalist Programme.

Though the three hypotheses mentioned above address different important issues, arguments that HCF have offered

---

4. N. Chomsky, *Rules and Representations*, London: Basil Blackwell, 1980.
5. N. Chomsky, "Language and Nature", *Mind*, January, 1995.
6. N. Chomsky, *New Horizons in the Study of Language and Mind*, Cambridge: Cambridge University Press, 2000.
7. N. Chomsky, "Language and Its Design: The Delhi Lecture, January 1996", in N. Mukherji, B.N. Patnaik, R.K. Agnihotri (eds.), *The Architecture of Language*, New Delhi: Oxford University Press, 2000. See also S. Pinker, *The Language Instinct*, New York: HarperCollins, 1995, for an interesting counter point.

for them are closely linked. The dual aspect hypothesis makes it evident that issues centring language cannot be addressed by sticking to the disciplinary divisions. But HCF have repeatedly maintained that the respective contents of FLN and FLB are to be empirically determined. Since FLN is both unique to language and unique to the human race, whatever aspects are shared with non-humans and/or evolved for purposes other than linguistic cannot constitute the content of FLN. On these grounds they have garnered empirical evidence to establish that neither the sensory-motor system nor the conceptual-intentional system can form part of FLN.

It is commonly held "Speech is Special" (vide Liberman and Leiberman) to the human species.[8, 9, 10] HCF point out that neither speech perception nor speech production is uniquely human. It was found that human infants, quite early in their life, are quite adept in categorical phoneme perception. In 1970, it was discovered that even chinchillas make similar discriminations. HCF cite three other findings: (a) that some animals can make distinctions based on formant frequency, (b) that tamarin monkeys can learn to discriminate the gross rhythm of different languages, and (c) that monkeys can perceive formants in their own species' vocalizations.

In respect of speech production, evidences revealed that both song birds and monkeys can produce species-specific formants. Though in non-human primates species typical vocalization is innate, in song birds this trait is acquired. Like humans, if song birds are deprived of the experience of

8. A. Liberman, *Speech: A Special Code*, Cambridge, MA: MIT Press, 1996.

9. P. Leiberman, *On the Origins of Language*, New York: Macmillan, 1975.

10. P. Leiberman, *Uniquely Human*, Cambridge, MA: Harvard University Press, 1991.

listening to their conspecifics, they develop highly aberrant song. Just as human adults find it difficult to acquire a new language, similarly many song birds tend to produce defective songs after a critical period in development. Young birds also pass through a phase of "subsongs" comparable to the "babbling" phase of human infants. These phenomena suggest that some aspects of speech perception and speech production were present long before the advent of language in humans. Maybe, these were related to the primitive vertebrate characteristics that evolved for general auditory processing than speech processing.

Speech production, especially acquiring a shared lexicon, also depends on learning by vocal imitation. Again HCF point out that vocal imitation and learning are not uniquely human. Dolphins and songbirds exhibit rich imitative capacities though surprisingly it is almost absent in monkeys and apes. Vocal imitation, thus proved to be an analogous trait, is relegated to the broad linguistic faculty by HCF.

Like the sensory-motor system, the conceptual-intentional system is neither uniquely human nor a necessary part of language. HCF provide evidence to demonstrate that the human conceptual system is present, though in a rudimentary form, in other primates. Chimp's and monkey's behaviour suggests that they engage in spatial, causal and social reasoning, though they cannot talk. Some have also ascribed mind-reading abilities to chimps. Vervet monkey's specialized calls show that they are able to use these as referential signals. That is why they can associate certain calls with certain things, though cases of such thing-word association are very limited in number, confined to here and now and stereotypical. Human beings, on the other hand, associate words and expressions with virtually any concept they can entertain. HCF, therefore, argue that conceptual and semantic abilities have very weak analogues and homologues in animal communication system. HCF concede that if strong evidences are available, they will declare this component of FLB as uniquely human. However, they do not find currently available evidences strong enough to consider C-I system either uniquely linguistic or uniquely human.

Let us now concentrate on the argument that HCF put forward in favour of their "recursion only" hypothesis. According to all linguists, the narrow syntax of any human language is recursive in nature. Recursion refers to a procedure that calls attention to itself or to a constituent that contains a constituent of the same kind. Thus iterability is an essential feature of recursion. Essential to recursive process is also the capacity to embed one unit into another. This recursive syntax lies at the root of limitless expressive power of human language. FLN takes a finite set of elements and yields a potentially infinite array of discrete expressions. Each of these discrete expressions is then passed to the sensory-motor and conceptual-intentional systems, which process and elaborate this information to pair each sound with at least one meaning. Chomsky is still thinking here in the model of a formal language, as he insists that structures are generated first while sound-meaning connection is produced later.

All natural language users are familiar with this property of discrete infinity. We can parse each natural language sentence into discrete units. Each sentence, for example, contains 6 words or 7 words but never 6.5 words. Because of the iterability feature, there is no non-arbitrary upper bound to sentence length. Language is similar to natural numbers in this respect. Some organism-internal factors that fall outside the scope of FLN and FLB curb the potential infinite power of FLN. For example, lung capacity imposes limits on the length of actual spoken sentences; working memory capacity restricts the complexity of sentences making them understandable. Other limitations that impose restriction on concept formation or motor output speed belong to FLB. HCF hold that it is possible to develop the evolutionary story of FLN without taking into consideration the constraints on FLB.

HCF insist that FLN does not have any analogue in the animal communication system and from this premise they arrive at the most controversial conclusion that FLN did not evolve through adaptation for communication, rather FLN evolved for reasons other than communication. Chomsky was of the opinion for a long time that "Language use is largely for oneself; inner speech for adults, monologues for children."[11]

> Language is not properly regarded as a system of communication. It is a system for expressing thought, something quite different. It can of course be used for communication as can anything people do — manner of walking or style of clothes or hair, for example. But in any useful sense of the term, communication is not the function of language, and may even be of no unique significance for understanding the functions and nature of language.[12]

What did then the narrow language faculty evolve for? HCF answer that FLN originated for some other narrow cognitive ability. "Specifically, recursion could have evolved in other animals to solve other computational problems such as navigation, number quantification or social relationship."[13] HCF write, "we focus here on the nature of number representation and the rule learning in non-human animals and human infants, both of which can be investigated

11. N. Chomsky, *On Nature and Language*, New York: Cambridge University Press, 2000, p. 77.
12. Ibid., p. 75.
13. M.D. Hauser, op. cit., p. 1578.

independently of communication."[14] Hauser and Fitch have experimentally established that animals and humans (both infants and adults) can represent number approximately as a magnitude with scalar variability. It is assumed that representation of number is required for object tracking, which is an important survival tool. It has, however, been established that animals fail to create open-ended generative systems. Chimps require thousands of training trials and often years to acquire the integer list up to nine. A human child, who has acquired the numbers 1, 2, and 3 and sometimes 4, grasps the successor function and can easily generate hundreds of numbers in sequence. Chimps, on the other hand, learn each numbers anew. A child's grasp of the successor operation is accompanied by "*ahā*" experience, which is totally absent in case of non-human primates. A human child learns an ordered list of symbols first and learns their meaning later. The reverse is the case of apes and parrots. They can be taught the meanings of numbers one by one without learning the list first.

Natural languages being recursive always have embedding of phrases within phrases, the handling of which requires at least a phrase-structure grammar. Hauser and Fitch in a few recent studies have shown that tamarins can at the most master a finite state version of the phrase-structure grammar, where recursion does not play any role. On the basis of these findings, HCF narrate their version of the evolution of the faculty of language thus. Interface systems of FLB are evolved first, then evolved the computational system which linked them. The computational recursive system is tractable in the habit of optimal foraging of many non-human species, involving minimal distance, recall of locations searched and kinds of objects retrieved. This domain-specific modular system of

14. Ibid., p. 1576.

foraging became domain-general under the pressure of evolution. HCF think it more likely that the domain-general ability as manifested through human language developed as a by-product of other kinds of neural organization than it being naturally selected for enhanced communication.[15]

HCF's story has been challenged most systematically and vehemently by Pinker and Jackendoff (hereafter PJ).[16] PJ appreciate the interdisciplinary move initiated by HCF and also admire FH's[17] endeavour to verify their hypotheses through well-designed and well-conducted experiments. Yet they argued that none of the three hypotheses is warranted by empirical evidence. PJ complain that the criterion of FLN/FLB distinction is not at all clear. HCF place all aspects that are not uniquely human within FLB. But it is not intelligible at all what they mean by uniquely human. In an evolutionary story some new trait can at best emerge from some older trait but nothing is *sui generis*. Besides, according to HCF's own admission, even recursion as such is not uniquely human, since recursive ability was present in different cognitive abilities of some non-human primates. But they are vehemently opposed to the story of the evolution of the language faculty through adaptation by natural selection. They rather prefer, as we have already pointed out, the by-product story *a la* Gould. But what is the advantage of the by-product story over the adaptation story? That is a question we should seriously apply our minds to.

---

15. W.T. Fitch, M.D. Hauser, Noam Chomsky, "The Evolution of the Language Faculty: Clarifications and Implications", *Cognition*, 97, 2005.

16. S. Pinker, R. Jackendoff, "The Faculty of Language: What's Special About It?", *Cognition*, 95, 2005.

17. W.T. Fitch, M.D. Hauser, "Computational Constraints on Syntactic Processing in Nonhuman Primates", *Science*, 303, pp. 377-80.

The next question that PJ raises is: Were Hauser and Fitch amassing evidence in favour of Chomsky's latest position called the minimalist programme (MP)?[18] The MP thesis recommends streamlining of the core of language systems. Since language is a mapping between sounds and meanings, only representations of sound (PF) and representations of meaning (LF) are really indispensable. For recursive combination of any two units, Chomsky admits one single operator in MP called Merge. He also admits a large number of principles of economy to keep in check construction of erroneous derivations by using Merge, e.g. certain operations are to be executed later rather than earlier, simple operations are to be preferred rather than complex operations, etc. MP thus cuts off all flabbiness present in previous versions of generative grammars, viz. the Extended Standard Theory, the Government-Binding Theory, etc. It is conceivable that this modification is effected by a single genetic change through drift or random process. Since Merge would be the only thing that had to be added to the pre-existing auditory, vocal and conceptual systems, invocation of natural selections to explain the adaptive complexity becomes unnecessary.

But this conjecture, point out PJ, has many difficulties. First, the minimalist syntax is not really minimalist. A simple sentence like "John saw Mary" has a tree with six levels of embeddings, four traces and five alternative derivations that need to be compared to ensure that one of the economy requirements has been satisfied. HCF consider language similar to invented formal symbol systems. Therefore, MP is for non-redundancy but in nature there is redundancy all around. So in a natural language, four different devices are often deployed redundantly. These are: phrase structure, linear order,

18. S. Pinker, op. cit.

agreement, and case. PJ, therefore, rightly point out, "Languages develop spontaneously in a community, subject to the vagaries of history, rather than being stipulated by formal arbiters."[19] PJ draw our attention to another lacuna in HCF's account. According to HCF, the recursive core of language is characterized by discrete infinity, but animal navigation systems show no such property. Besides, wherever recursive number system developed, language had developed prior to that. So it is more likely that the cognition of recursive number has been exapted from the recursive principles of language and not vice versa. Besides, mathematical recursion is different from recursion required for object tracking and foraging. In response to PJ's objections HCF have categorically denied any connection between the minimalist programme and the exaptation thesis on FLN, yet I do think that PJ's suspicion is well founded. It is PJ's considered opinion that language needs to be recursive because its function is to express recursive thoughts. Our conceptual-intentional systems develop infinite concepts and relate innumerable words to things via such concepts. To organize those concepts we need recursive syntax. If communication were not the purpose of the evolution of the faculty of language, then all attempts at establishing semantic relations also become superfluous. Chomsky, since 1975, upholds that the innate aspect of the language faculty is for learning language from the community and not for inventing language. Nobody takes so much trouble of developing a complex syntactic system only to talk to oneself. A system for talking to oneself need not develop tools for speech production. But the fact that phonemic tools have been developed in course of evolution increases the plausibility of the adaptive story. HCF theses are not well substantiated by

19. S. Pinker, op. cit., p. 226.

empirical evidence or linguistic fact. Still Chomsky will go undaunted. Probably, he will not consider this a real refutation at all. Maybe, it is merely an apparent refutation, just as the phenomenon of the world appeared to refute the Copernican thesis.

HCF's story of the language faculty appears implausible because of the fact that it is an add-on to Chomsky's earlier theories. Over more than fifty years Chomsky has developed a theory of human language against the nativist backdrop of the Neo-Cartesian philosophy. But when he tried to conjoin whatever insights he had about the human language faculty with a viable theory of evolution, two stories did not gel properly. Chomsky has always maintained that a human child is born knowing certain facts about universal constraints on possible human languages. It is the integration of this innate knowledge with a corpus of primary linguistic data, derived from the child's observation of the adult linguistic behaviour of the speech community that explains the eventual assimilation of mature linguistic capacities. The interaction between the innate structure and the primary linguistic data is computational. The notion of computation has all along been connected with certain logico-semantic relation among symbols. Here I would like to make two observations. His Cartesian propensity may be behind his search for something uniquely human in the core of the language faculty. His distinction between the core and periphery of the faculty of language is once again an inspiration of his adhering to the Turing-Neumann model of computation where peripheral subsystems are controlled by a central executive system and the input-output system is always mediated by a rule-governed representational system. To a cognitivist, this layer of representation is all-important where semantics is preserved through syntactic manipulation of symbols. It appears that

HCF could not go beyond this legacy. To me, their exaptation theory also appears somewhat artificial. In the pre-human stages of evolution HCF admit some domain-specific faculties, e.g. the number faculty which subsequently produced as a by-product the faculty of language. Sure, there are instances of new functions being ascribed to old body parts in course of adaptation to the environment. But we have not yet heard that any abstract mechanism like the computational mechanism devoted to one function has been exapted to serve some other function. Besides, one may raise the question whether animal foraging and navigating systems are at all computational. Recent experiments on animals have been viably interpreted in terms of distributed cognition where use of representation is considered superfluous for explaining simple foraging and navigating behaviour of animals. It is possible to describe such behaviours of animals in computational terms but it is like describing the planetary motion in terms of Kepler's law. Planetary motion at best conforms to Kepler's law, planets do not move in accordance with it.

It is true that recursion which is definitely an important process of human languages was not found in the animal communication system. But what is the possible reason for that? To answer that question we have to give up the view-points of the first and the second generation cognitive scientists and adopt the dynamic systems view-point of the third generation cognitive scientists. We have to remember that when a trait, unique to a particular species comes into existence in course of evolution, the inner structure of an animal species as well as its environment are responsible for that. Neither HCF nor PJ viewed the facts from this perspective. Both viewed language evolution as an abstract problem (PJ's account is not totally ahistoric) and not as a dynamic process that occurred at a specific time, in a specific location as merely a part of the overall evolution of the human species. Animal communication systems were not characterized by recursion because these did not possess adequate units on which recursion can apply. ACS units are, we have seen, indexical, confined to here and now. These units are associated with stereotypical meaning. Two calls of vervet monkeys can be conjoined but one cannot be embedded in the other. It was neither necessary nor possible. Human beings started using linguistic symbols in the true sense of the term by associating signs with remote objects. So units of human language are characterized by symbolization, predication and displacement. Though primate communication systems lack these traits, these are present in the communication systems of bees and ants in rudimentary form. Why should these primitive organisms possess capacities that even the great apes lack? Surely it was not due to the lack of neural infrastructure but something else. What is that? Ants and bees had motivations for developing symbol systems which the primates lacked. Most probably, the standard indexical communication system was inadequate for aerial foraging of bees and terrestrial foraging of ants. They thus developed rudimentary symbol systems for communication. Now we must remember what the third generation cognitive scientists pointed out umpteen times — that evolution is not a one-way process in which a largely variable but autonomous environment selects the population of living organisms, as if the role of the organism in evolution is essentially passive. That is not true. The environment, to a large extent, is the creation of the organism that inhabits it. Since the niche forms the organism just as much as the organism forms the niche, we think that difference in non-primate and primate niches is the obvious place to look if we are to understand why language appears to develop all on a sudden in *homo sapiens*, while it was absent in their ancestors.

Let me end this paper by summing up the points that I want to make.

- The faculty of language evolved for communication.
- Symbolization, predication and displacement constitute the bases of human language. Humans got the hang of recursion only after they developed recursion-friendly units of language.
- We need not subscribe to the exaptation hypothesis for explaining how human language developed in the absence of homologous precursors. A two-way theory of evolution will probably lead us to the correct solution of the problem.

14

Content Externalism versus Eliminative Materialism
A Case against Scientism

Manidipa Sen

ELIMINATIVE materialism with regard to intentional mental states is an influential thesis in analytical philosophy. The main aim of this paper is to review a set of arguments against eliminativism, particularly eliminativism of the kind developed by Paul Churchland, advanced by Gregory McCulloch. The paper shall try to establish a form of content externalism as against eliminativism, and shall attempt to show that eliminativists, very much in line with the Cartesians, draw a distinction between the subjective and the objective which is untenable. Underlying the project of eliminativism is a strong form of naturalism which also paves the way for a faulty conception of both the subjective and the objective, and thus needs to be abandoned. The conclusions that the paper shall try to establish, following McCulloch, are firstly, that the subjective does not exclude the objective, and secondly that we still need to maintain a distinction between knowing the mental and knowing the physical.

I

Let us begin at the very beginning by stating certain preliminaries.

Questions regarding the nature of mind and of mental states, the relationship between the mind and the external world, the exact connection between language and mind in linguistic expressions/articulations of mental states and processes are some of the basic problems that have intrigued philosophers throughout the centuries. How do speakers express their mental states directed towards the world through the use of language is a question concerning the relationship between mind, language, and the world. We should however note that these issues in philosophy of mind have become extremely important in analytical philosophy in recent times. Though analytical philosophy started with what is known as the "linguistic turn," we can say that a "mentalistic turn" has taken place in the later part of the development of analytical philosophy.[1] This movement, it may be argued, can be traced back to Frege's distinction between sense and reference and the relationship between the two. Many philosophers, for example Evans, McDowell, McCulloch, etc. have tried to show the relationship between this Fregean distinction in the domain of meaning and the mentalistic turn in analytical philosophy. For instance McCulloch says, "theory of *Sinn* embraces mind-language relations, theory of *Bedeutung* embraces language-world relations. Somehow, then the two together embrace *mind-world relations*, in many ways the core of philosophy, and the heart of the Cartesian tradition."[2] In fact, one may very well say that some of the recent debates in the philosophy of mind centering around the internalist/externalist debate can be traced back to Frege's theory of meaning and cognitive

1. For detailed discussion see Manidipa Sen, *Thinking about the World: An Essay in De Re Thoughts and the Externalist/Internalist Debate*, Shimla: Indian Institute of Advance Study, 2008, especially the Introduction.

2. G. McCulloch, *The Mind and Its World*, London: Routledge, 1995, p. 42.

significance. These three parameters of human thinking: mind, language and the world, and the relationship between the three is traditionally encapsulated in the notion of intentionality, the fact that mental phenomena involve what Brentano called "a direction upon an object," and what contemporary philosophers call "aboutness'' or "world directedness." Along with the essential intentionality of the mental, we shall here assume the following characteristics about the mind as essentially embodied, and hence cannot be studied in isolation from the body; as essentially interactive, and hence is constantly negotiating with other minds and the external world; as essentially a "being-in-the-world," and hence is on a par with other things in the world.

As we may infer from these characteristics, the picture of mind summed up in these few characteristics is anti-Cartesian and is essentially opposed to the view that since mind and world are two independent categories where the world has no constitutive role in the individuation of mental states and hence to have a full understanding of such states or to individuate them it is sufficient to know what goes on inside the head. This is the main point of the famous internalist/ externalist debate in the philosophy of mind.

This controversy has been developed in detail in analytical philosophy. Hilary Putnam's path-breaking paper "The Meaning of Meaning," provides us with an externalist account regarding mental content.[3] On the contrary, Fodor's idea of methodological solipsism in the philosophy of mind (developed in his writings like *Psychosemantics*) is a reassertion of the Cartesian internalist model of the mental.[4]

---

3. H. Putnam, "The Meaning of Meaning", in his *Mind, Language and Reality*, vol. 2, Cambridge: Cambridge University Press, 1975.
4. J. Fodor, *Psychosemantics*, Cambridge, MA: MIT Press, 1987.

II

The fact that there are intentional mental states that are object-directed is accepted from a common sensical point of view. Taking this to be a kind of theory[5] providing us with an adequate explanation of our everyday psychological understanding, one may ask further: Is it really a good theory about human mind? Is it able to account for the beliefs, desires causing actions in accordance with certain general principles concerning their relationship? According to many scientifically oriented philosophers, our common sense conception of psychological phenomena, though wants to provide us with a theory explaining mental phenomena, this theory can be shown to be radically false. As opposed to this theory, therefore, what we require is scientific psychology. Given the fact that common sense psychology and scientific psychology talk about the same thing — the mind — one may legitimately ask: what constitutes this opposition? There are many different kinds of answers provided by philosophers to this question. But they may broadly be divided into two kinds: (1) *Justify or vindicate* the common sense generalizations, which are taken to be largely true, in terms of scientific psychology; or (2) since the common sense theory is largely false, it is important that a science of the mind be developed without having to employ the categories of common sense psychology, and in the process common sense psychology shall be *eliminated* in favour of scientific psychology. The second is the kind of approach developed by Paul Churchland in many of his writings, particularly, in the book, *Scientific Realism and the Plasticity of*

5. Common sense psychology is assumed to consist of both generalizations (or laws) and specific theoretical posits, denoted by our everyday psychological terms like "belief" or "pain." The generalizations are assumed to describe the various causal or counterfactual relations and regularities of the posits.

the Mind, and in his article, "Eliminative Materialism and the Propositional Attitudes."[6, 7] He starts the article by defining eliminative materialism thus:

> Eliminative materialism is the thesis that our common sense conception of psychological phenomena constitutes a radically false theory, a theory so fundamentally defective that both the principles and the ontology of that theory will eventually be displaced, rather than smoothly reduced, by completed neuroscience.[8]

Eliminative materialists argue that the central tenets of common sense psychology radically misdescribe cognitive processes; consequently, the intentional states with which common sense psychology is concerned pick out nothing that is real.

We may note here that like dualists, eliminative materialists insist that mental states ordinarily understood cannot in any way be reduced to or identified with neurological events or processes. However, unlike dualists, eliminativists claim that there is nothing more to the mind than what occurs in the brain/neurophysiology. The reason mental states are irreducible is not because they are non-physical; rather, it is because mental states, as described by common sense psychology, do not really exist. That is why we need to get rid of common sense psychology altogether.

The position of the eliminative materialists may be understood better if we realize that their point is not merely

---

6. P. Churchland, *Scientific Realism and Plasticity of the Mind,* Cambridge: Cambridge University Press, 1979.
7. Paul Churchland, "Eliminative Materialism and Propositional Attitudes", *Journal of Philosophy,* 78, 1981; reprinted in William Lycan (ed.), *Mind and Cognition,* Oxford: Blackwell, 1999.
8. Ibid., p. 120.

to relocate entities of the replaced theory into the new theory that is proposed. It is not a mere revisionist project. It is a far more radical ontological position which argues that there is nothing in the common sense psychology which needs to be relocated in the framework of scientific psychology. Various common sense psychological concepts like the concept of belief will eventually be recognized as empty concept that fails to correspond to anything that actually exists.[9] So to sum up:

> Scientific psychology has no obligation to explain why the common sense generalizations are true, because there are good reasons to think that they aren't true! So we should expect scientific psychology eventually to eliminate common sense rather than to vindicate it.[10]

---

9. Quine also endorsed the idea that mental notions like belief or sensation could simply be abandoned in favour of a more accurate physiological account. In a brief passage in *Word and Object*, Cambridge: M.I.T. Press (1960), Quine suggests that terms denoting the physical correlates of mental states will be more useful and, as he puts it, "[t]he bodily states exist anyway; why add the others?" (p. 264). However, Quine goes on to question just how radical an eliminativist form of materialism would actually be, implying no significant difference between explicating mental states as physiological states, and eliminating mental state terms in favor of physical state terms. He asks, "Is physicalism a repudiation of mental objects after all, or a theory of them? Does it repudiate the mental state of pain or anger in favour of its physical concomitant, or does it identify the mental state with a state of the physical organism (and so a state of the physical organism with the mental state)" (p. 265)? Quine answers this question by rejecting it, suggesting there is no interesting difference between the two cases: "Some may therefore find comfort in reflecting that the distinction between an eliminative and an explicative physicalism is unreal" (p. 265).

10. T. Crane, *The Mechanical Mind*, London: Penguin Books, 1995, p. 72.

What are the arguments put forward by eliminativists in favour of this view? One set of arguments, found in Churchland's paper, "Eliminative Materialism and the Propositional Attitudes,"[11] is negative in nature. Their main thrust is to show why common sense psychology as a theory is unacceptable. After having established the claim that common sense psychology is a theory, Churchland goes on to show why its "principles are radically false and its ontology is an illusion"[12] by highlighting three deficiencies of common sense psychology. The three deficiencies are: explanatory inadequacy, stagnation and finally incoherency and discontinuity with "other well-established theories in adjacent and overlapping domains."[13] None of these arguments against common sense psychology are that convincing. I shall not go into the details of the problems with these arguments[14] because that is not our direct concern since our main task here is not to defend common sense/folk psychology. We shall, however, try to show what is particularly inconsistent with Churchland's position.

III

Eliminative materialism, therefore, is a philosophical position which holds that common sense is just mistaken in supposing that people have beliefs, desires, hopes, etc. that is, they have some form of intentionality. Common sense is most probably mistaken also in supposing that people have sensations and feelings as we think that they have.

---

11. Paul Churchland, "Eliminative Materialism and Propositional Attitudes", *Journal of Philosophy*, 78, 1981; reprinted in William Lycan (ed.), *Mind and Cognition*, Oxford: Blackwell, 1999.
12. Ibid., p. 122.
13. Ibid.
14. For a detailed discussion and critique of these negative arguments put forward by Churchland, see note 10, pp. 70-76.

We may, at this juncture, note the motivations which may lead to a theory of this kind.

1. One of the main motivations for these philosophers, along with most of the philosophers of mind of this age, is the rejection of immaterialism. So, for these philosophers, the main task of philosophy of mind is to replace any form of immaterialism in favour of the materiality of the mind. About this move, McCulloch[15] says that this first step towards a materiality of the mind does not solve many of the problems that arise from Cartesian dualism. The question as to how the material mind relates to the environment (both social and physical) in general and to the body in particular is still a pending question which is not automatically taken care of by establishing a materialist conception of the mind. There are a whole range of metaphysical issues which remain unresolved even after immaterialism is rejected.

2. Another important motivation for a view of the above kind is a form of scientism. This is roughly the position which treats science as the only source of knowledge. But one cannot take this position seriously unless one takes the scientific talk of a wide variety of unobservable entities seriously.[16] This, in turn, implies the existence of the world as it is regardless of how it impinges upon our awareness. Thus this scientism leads to a form of scientific realism which in a way supersedes logical positivism as a dominant ideology of analytical philosophy. That is why, we see Churchland raising the objection against common

15. G. McCulloch, *The Mind and Its World*, pp. 7-9.

16. Ibid., p. 113.

sense psychology that it fails to synthesize with "theories about adjacent subject matters."[17] He goes on to say,

> If we approach *Homo sapiens* from the perspective of natural history and the physical sciences, we can tell a coherent story . . . which encompasses particle physics, atomic and molecular theory, organic chemistry, evolutionary theory, biology, physiology and materialistic neuroscience. That story, though still radically incomplete, is already extremely powerful, outperforming F[olk] P[sychology] at many points even in its own domain.[18]

This is surely an extreme form of scientism which claims that for any theory to be respectable it has to be integrated into a broader framework of science. But, though science talks about and admits of a whole range of unobservables, it cannot talk about or admit of beliefs and desires. This is a point which shall be taken up for further scrutiny in the forthcoming section. For the moment, one may mention one misgiving regarding this line of thinking. In order to go through the elimination properly we really need to change ourselves, and get out of the habit of thinking in terms of beliefs and desires altogether, or to put it more bluntly "stop seeing other humans as persons."[19] There may be a real issue over common sense psychology — regarding whether we can take it to be

17. Paul Churchland, "Eliminative Materialism and Propositional Attitudes", *Journal of Philosophy*, 78, 1981; reprinted in William Lycan (ed.), *Mind and Cognition*, Oxford: Blackwell, 1999, p. 124.

18. Ibid.

19. G. McCulloch, *The Mind and Its World*, p. 220.

a proper theory at all. But all the arguments against our common sense understanding of human mind are coming from, to quote McCulloch, "the tendency to accord exaggerated respect to the scope and power of natural science. Analytical Philosophy, especially Philosophy of Mind, is currently awash with scientism."[20]

IV

Greg McCulloch in his posthumously published book, *The Life of the Mind: An Essay in Phenomenological Externalism,* tries to show that Churchland's position *vis-à-vis* the mental and the physical found in his account of eliminative materialism is deeply incoherent. According to McCulloch, the underlying problem with this kind of scientific realism is its adherence to a form of naturalism which tries to "model the mind on indisputably physical or mechanical processes, without remainder."[21]

We may, at the very outset, try to articulate the incoherence that McCulloch sees in Churchland-style eliminativism, and then try to work out what leads to this type of incoherence. According to eliminativists like Paul Churchland our world-view off and on needs to be reordered in accordance with the best available scientific theory. This kind of reordering may be called "W-recalibration."[22] Furthermore, in view of the new scientific theory we also need to reorder our view of the mind and the mental, or what is

---

20. G. McCulloch, op. cit., p. 220.

21. G. McCulloch, *The Life of the Mind: An Essay on Phenomenological Externalism,* London: Routledge, 2003, see p. 55.

22. Ibid.

called "M-recalibration" by McCulloch.[23] This reordering has to be comprehensive, such that we change the way in which we appear to ourselves as well as to each other as minded beings. But McCulloch argues that holding on to these two forms of recalibration leads to a deep incoherence in Churchland's position due to the fact that "Churchland is an externalist about the mind when W-recalibration is the issue and anti-externalist when M-recalibration is"[24] which he can't be.

McCulloch traces back the reason for this sort of incoherence to two conceptions of the subjective that Churchland works with: one where "subjective states such as seeing that . . . have objective intentionality" and hence are in some sense externally individuated, another where the notion of subjective is taken in a traditional Cartesian way where "subjective excludes the objective"[25] altogether. Our main task in this paper is to show how McCulloch justifies this claim of incoherence. In the process a more radical notion of the subjective as well as externalism are likely to emerge that may have more important consequences on philosophy of mind in general.

Let us first turn to Churchland's idea of W-recalibration. What does it really mean when we say that our world-view needs to be reordered in accordance with the available scientific theory? What happens when a group of people recalibrated as they are "do not observe the Western sky redden as the sun sets?"[26] They, according to eliminativists, "observe the wavelength distribution of incoming solar

23. G. McCulloch, *The Life of the Mind*, p. 56.

24. Ibid.

25. Ibid., p. 66.

26. P. Churchland, *Scientific Realism and Plasticity of the Mind*, p. 29.

radiation shift towards the longer wavelength, etc."[27] So, the point about recalibration of our experience about the external world is that we are able to replace our ordinary perceptual experiences and hence perceptual judgements by whatever is given to us by the available scientific theory. From this it is obvious that our perceptual experiences and perceptual judgements about the external world cannot be determined in a theory-neutral way and to that extent they cannot be assigned any privileged status. Hence our perceptual experience and perceptual judgement about the world in this picture of recalibration have to be determined externally.

A few points about this position should be highlighted at this juncture:

1. W-recalibration, according to McCulloch, takes place in and through a "self-realignment."[28] It is not that external observers re-configure our experience so that they be on a par with true scientific theory. We ourselves have to modify our conceptions of things. "It is a matter of self-interpretation or self-reinterpretation, in the light of our best account of how the world stands."[29]
2. Churchland's proposal for W-recalibration is, however, not an appeal that ordinary human beings should learn a lot of science. It is much stronger than holding that we have science-based beliefs about the surrounding world. He denies that we may think as scientists and perceive like ordinary people. According to McCulloch,

---

27. P. Churchland, *Scientific Realism and Plasticity of the Mind*, p. 29.
28. G. McCulloch, *The Life of the Mind*, p. 57.
29. Ibid.

Churchland's idea is that our "contentful perceptual states, seeing that this, hearing that the other, have content which are imbued with concepts shaped by the appropriate scientific theory."[30]

3. Furthermore, Churchland's proposal of W-recalibration definitely implies that the meaning of observation terms cannot be given in sensation. Observation terms involve objective intentionality, which are articulated in terms of their causes. And it is science which tells us "what are in fact the typical cause of our 'sensations,' so tells us what they are objectively about."[31] In this way our beliefs about them are regulated by the best available scientific account of them.

What emerges through this short account of eliminative materialism of the kind upheld by Churchland is that to say that we have contentful perceptual states is to admit that these states are embedded in objective theoretical framework provided by science and thus are to be identified in terms of that theoretical framework. In this respect "the intrinsic qualitative identity of one's sensations is irrelevant to what properties one can or does perceive the world as displaying."[32] So one may say that Churchland is arguing for externalism regarding the content of our perceptual states by invoking objective intentionality which is in fact strengthened by his scientific realistic framework. This, in turn, implies that when it comes to W-recalibration Churchland does not accept any purely subjective domain of our experience that is experience about the world is interpenetrated by objective world of the scientific theory,

30. G. McCulloch, *The Life of the Mind*, p. 57.

31. Ibid., p. 59.

32. P. Churchland, *Scientific Realism and Plasticity of the Mind*, p. 15.

and has to do with objective intentionality. It is a misrepresentation of our cognitive nature to claim that our perceptual content is given in sensation. A complete account of our conception of things is carried out in accordance with what our best scientific theories tell us about the world.

What happens in case of M-recalibration? What happens when we want to provide a theory of our beliefs and desires? This point of Churchland is, by now, quite well known. His "science-driven" arguments against common sense psychology not only lead to re-ordering/recalibrating our talks about the mental, it ultimately leads to an altogether elimination of these mental categories. As McCulloch rightly points out,

> a further application of science, this time in the intentional realm is supposed to point the way to a further dose of recalibration, not with respect to the world we confront but with respect to ourselves . . . and perhaps beyond that elimination altogether.[33]

M-recalibration cannot be separated from W-recalibration. Since W-recalibration changes our view about the external world, it also involves changing our view about ourselves as being part of the external world. It implies that a better description of ourselves should involve something other than beliefs and desires, about which no scientific theory is available. As McCulloch says, according to Churchland, "we shall find that 'belief' and 'desire' don't effect useful classifications at all, and are simply philosophers' abstractions."[34]

What leads Churchland to eliminate mental states of beliefs and desires in terms of neuroscience? As we have mentioned

33. G. McCulloch, *The Life of the Mind*, p. 64.

34. Ibid.

earlier, it has to do intrinsically with his reservations concerning common sense psychology as a theory about the mind. At this point, according to McCulloch, Churchland "collapses two claims into one; the claims that our self-conception is part of (a) a theory which is (b) *about our inner workings*."[35] But this way of characterization of self-conception as something to do exclusively with inner working is obviously anti-externalistic. Hence McCulloch rightly remarks "claims about intentional matters are here construed as claims about things which take place exclusively inside us, just as, presumably, neurophysiological happenings do."[36] Therefore, it seems that we cannot hold on to W-recalibration along with eliminativism. So, either Churchland has to reject externalism or reject eliminativism. Rejection of the former, for him, amounts to a rejection of the scientific realistic stance, whereas rejection of the latter amounts to an acceptance of an inner realm of beliefs and desires dissociated from the world.

V

What conclusions about the mind as well as about the world may be drawn from the eliminative materialist picture of W-recalibration and M-recalibration? The picture is quite simply this. Our external environment is fully determined by the best available scientific theory which tells us what really is there in the world. So, the objective world as well as "objective intentionality" is devoid of any subjective element. Regarding the mental, the point of view is that, if there is to be a mental domain at all, it has to be a domain of inner realm devoid of any external interference. And McCulloch's more general criticism against this position is that "it draws the line

35. G. McCulloch, *The Life of the Mind*, p. 65.

36. Ibid.

between the subjective and the objective in the wrong way."[37] Whereas according to McCulloch, "the subjective does not exclude the objective: much less is it inside our heads; much less is the objective 'out there.'"[38] So, the picture of the mind that McCulloch wants to draw is a picture where mind is not a passive spectator of what is going on before the mind. In this context one would like to recall the arguments put forward by Davidson in his paper, "The Myth of the Subjective."[39] According to both Davidson and McCulloch, the picture of the subjective that both the extreme materialists and the extreme mentalists have are really faulty. The main task for a philosopher of mind is to think of mind and world relationship outside of the inner-outer metaphor. In a recent paper this has been elaborated upon, and so we shall leave it at that.[40]

Nonetheless both McCulloch and Davidson are in favour of externalism. Externalism of some form or the other (be it social or environmental) regarding mental content is also a position widely accepted by philosophers of mind in recent times; in fact, according to McCulloch, it has somewhat become an orthodoxy in analytical philosophy. So, the obvious question that arises in one's mind is whether there is anything special about and, if so, what is so special about McCulloch's form of externalism. McCulloch, as Tim Crane, in the introduction to McCulloch's book, *The Life of the Mind,* quite rightly points out

37. G. McCulloch, *The Mind and Its World,* p. 54.

38. G. McCulloch, *The Life of the Mind,* pp. 54-55.

39. D. Davidson, *Subjective, Intersubjective, Objective,* Oxford: Clarendon Press, 2001.

40. M. Sen, "On the Myth of Subjective: A Wittgensteinian Perspective", *Journal of Interdisciplinary Crossroads* 3 (2006) (forthcoming).

that McCulloch is arguing for a radical form of externalism.[41] Hence one needs to see what this radical form of externalism is, what picture of "the life of the mind" that this form of externalism provides us with.

McCulloch argues in favour of "content externalism," which is a thesis regarding how mental states can be individuated in terms of the content to which the mental state is directed.[42] As opposed to the Cartesian position that the psychologically appropriate way of individuating types of mental states is by their internal/narrow content, "content externalism involves the claim that content or meaning is environmentally constrained."[43] Mental states, according to these philosophers, are to be understood in terms of some external environmental causes, since contents of such mental states are to be understood in terms of their external environmental causes too. Hence, instead of taking mental content to be determined entirely by what is going on in the head, it is essential to see what relation the subject has to the things of the world in order to determine the content of the subject's mental states.

In his attempt to argue for a radical externalistic position, McCulloch invokes the arguments put forward by Putnam in his famous article "The Meaning of 'Meaning,'" to show that meanings are not in the head. He also takes into consideration the arguments provided by McDowell to show that there can't be any *Sinn* (in Frege's sense) without *Bedeutung*. These two propositions, viz. (1) meaning is not in the head and (2) there can't be any *Sinn* without *Bedeutung*, together establish a form of content externalism. This, in turn, brings forth a conception

41. G. McCulloch, *The Life of the Mind*, p. xiii.

42. Ibid., Ch. 2.

43. Ibid., p. 41.

of intentionality where there cannot be any thought without there being a world towards which the thought is directed. As McCulloch succinctly puts it at the end of the second chapter, "There can be no subjectivity when all the objective is missing."[44] This position is entirely in opposition to the Cartesian ontological real distinction between the mind and the body which results in Descartes' position that mental states are identical if and only if they are introspectively indistinguishable. This mental-physical dichotomy in another form lurks behind the eliminativist claims as well.

What distinguishes McCulloch's form of externalism is the fact that it does not say that mental states are relational in nature in the sense that there is some intrinsic feature of the mental that is related essentially to something outside of it. As Tim Crane in the foreword to the book points out, this form of externalism "attempt(s) to give up this whole way of thinking in terms of intrinsic and extrinsic features of mind at all."[45] The reason being that for externalism of this variety, far from mind and world being metaphysically independent entities with mind being essentially related to the world, what we require is that a subject's subjectivity and embodiment not be abstracted from the world. Subjectivity is essentially being-in-the-world, and is phenomenologically constituted. This is where McCulloch's idea of content being phenomenological as well as externalistic needs highlighting.

Though a pure subjectivity is denied at the ontological domain, we can still maintain a distinction between the ways we come to know the mind from the way in which we come to know the physical world. So, from an epistemological point of view there is a distinction to be made between these two

44. G. McCulloch, *The Life of the Mind,* p. 53.

45. T. Crane, op. cit., p. xiii.

forms of knowledge. This is another way of attacking the eliminativist materialist position. Understanding minds is radically different from understanding objects of physical science. In the latter case we take recourse to causal laws, objective relations etc. whereas in the former case what we require are categories like "interpretation, reason, empathy." And the latter categories cannot be entirely replaced by or eliminated in favour of the former.

15

Empirical Knowledge and the Inner-Outer Distinction

Madhucchanda Sen

IN this paper I shall be discussing an age-old problem in philosophy, viz. how empirical knowledge is possible. It is yet another attempt to understand "the relation between the human mind and the rest of nature, the subjective and the objective as we have come to think of them."[1] In doing so I shall be trying to understand better and as a result loosen the hold on certain traditional distinctions in philosophy, viz.: "inner-outer," "internal-external," and "subjective-objective." These distinctions, though deeply entrenched in our tradition, have overburdened us with certain prejudices, which have created significant stalemates in metaphysics, epistemology, and philosophy of mind. In metaphysics this dualism has at the same time sustained and threatened realistic theories. In epistemology this dualism has made the threat of scepticism ever-present. In philosophy of mind this dualism has most notorious consequences, to name only a few: Cartesian Dualism, Problem of Mental Causation, the Problem regarding non-conceptual content of experience, the Externalism-Internalism Debate.

---

1. D. Davidson, "The Myth of the Subjective", in his *Subjective, Intersubjective, Objective*, Oxford: Clarendon Press, 2001, p. 39.

In discussing this distinction and the enormous consequences of it in epistemology and in philosophy of mind, I shall have John McDowell's *Mind and World* and the specific critique of this dualism as we find it in this book, as my backdrop. Following McDowell, I will try to loosen the hold on this traditional distinction. McDowell proposes a new way of looking at experience which makes the problem that emanates from the inner-outer (scheme-content) distinction dissolve. I shall discuss the way that McDowell tackles the problem of empirical knowledge and show how his solution has had lasting effect on philosophy of mind. I shall be discussing first the idea of non-conceptual content of perceptual experience which results from the inner-outer divide and how McDowell counters this idea. I shall then be discussing how McDowell's response to this inner-outer distinction also determines his stand on the externalism-internalism debate about mental content.

I

McDowell on the Myth of the Given

The question that we start with is: How do rational minds make contact with the world? In the empiricist tradition, it is believed that the gap between the mind and the world can be best met by sensory experience, which is the only thing that can possibly bridge the gap by providing evidence of the external world. The plot thickens as we find that there seems to be no acceptable conception of experience that can really do this bridging. The main problem that one faces when one tries to establish the possibility of empirical knowledge is, how can it be shown that our thoughts are rationally constrained by external reality. If our sensory experience is going to help us in meeting the sceptics challenge and in showing how the gap between the rational mind and the natural world can be bridged, then sensory experience has to meet two criteria: (1) Experience itself has to be appropriately constrained by external reality, i.e. it must be "receptive"; and (2) Sense experience must be able to enter into logical relation with belief, i.e. it must be "conceptual." Our empirical beliefs must be somehow constrained by features of the external environment via experience, otherwise those beliefs would be utterly divorced from the world that they are supposed to be about; yet our empirical beliefs must be somehow free of external control, else we could give no sense to the idea that we are responsible for *our* beliefs. We are definitely reminded of the puzzle that was created by the Kantian slogan, "thoughts without content are empty, intuitions without concepts are blind." But it seems to be impossible to unify the receptive and the conceptual. Wilfred Sellers in his "Empiricism and the Philosophy of Mind"[2] argues against this conception of experience that can perform this double task and terms it as "the Myth of the Given."

Traditionally, possibility of empirical knowledge has been attempted to be proved by empiricists by having recourse to the idea of "the Given." It was believed that in experience the world is revealed in its true guise, and it is this bare given which will prove to be handy when we try to come up with a justification of our empirical knowledge. This idea of "the Given" received severe criticism from two quarters. Conceptual relativists have always been suspicious about the possibility of the world being presented to us, as it is, without any intervention of the conceptual mechanism that helps us to

2. W. Sellars, "Empiricism and the Philosophy of Mind", in H. Feigl and M. Scriven (eds.), *Minnesota Studies in the Philosophy of Science,* Volume I: *The Foundations of Science and the Concepts of Psychology and Psychoanalysis,* Minneapolis: University of Minnesota Press, 1956, pp. 253-329.

understand it. Donald Davidson in his essay, "A Coherence Theory of Truth and Knowledge,"[3] criticizes the concept of "the Given," but his criticism is different from the one extended by conceptual relativists. Davidson says that traditional empiricism, in various forms of foundational theories, subscribes to a dualism between scheme and content and it is this dualism that leads to the idea of "the Given." Philosophers in their attempt at meeting sceptical challenges have usually spoken of perceptual experience where the world seems to be presented to us. It is usually believed that perception is the window to the world. This is why many empiricist philosophers have thought it to be absolutely necessary to come up with a theory of justification for empirical knowledge in terms of what is given or presented in our experiences. All other beliefs and claims to knowledge are justified in the light of these comparatively basic beliefs/experiences where the world is presented in its true guise. We find philosophers talk about sense data, about self-presenting states, about qualia, about the phenomenal character of experience — and they do so all in order to find some way of showing that there is an external check on our knowing mechanism which can work as a justification to our knowledge claims about the world. All these issues received comprehensive attention within the scope of philosophy of language in an essay by Donald Davidson, called "A Coherence Theory of Truth and Knowledge."

The scheme-content dualism and the concept of "the Given" that is generated by it received criticism from Davidson on two counts. First of all Davidson does not accept this dualism as it necessarily presupposed the idea of conceptual schemes which he finds unacceptable. Secondly, he is critical

3. D. Davidson, "A Coherence Theory of Truth and Knowledge", in, E. Le Pore (ed.), *Truth and Interpretation: Perspective on the Philosophy of Donald Davidson*, Oxford: Basil Blackwell, 1986

of the idea of "the Given" because he feels that the justificatory role that is assigned to "the Given" can never be assigned to it. Foundationalists believed that the world is *given* in its true guise in experience and hence all empirical beliefs have to finally gain justification from this bare given. Davidson shows within his coherence theory of justification that this given can at best be a causal intermediary and not an epistimic intermediary and hence cannot be assigned any glorious justificatory role.

John McDowell, in his book, *Mind and World,* revives the Kantian tradition in the wake of the criticisms brought against traditional empiricism on the issue of empirical knowledge. In doing this he is trying to show that we need not subscribe to conceptual relativism or to a coherentist theory (as Davidson does) and yet through a revival of the Kantianism, the tradition can be "exorcized" of the scheme-content dualism which gave rise to the myth of "the Given."

John McDowell is more gentle in his critique of this dualism and the idea of "the Given." First of all, McDowell points out that the Given, in spite of all the misgivings associated with it, provides us with a viable conception of experience which could bridge the gap between mind and the world. And if we summarily give up this notion then we will be faced with what may be called a blind internalism. The greatest problem with this kind of internalism is that it makes our beliefs not only lack justification but also content. Though McDowell is reviving a Kantian idea in order to bust the myth of the Given and the scheme-content dualism that nests it, he is critical of the Kantian idea of the unconceptualized world in itself. One might think that this would make his thesis slide into a kind of Idealism or empty internalism. But McDowell claims that this is not so, and that his view is a realistic view.

The main aim of McDowell's criticism is that the idea of

"the Given" and the role that is believed by traditional empiricists to have in justifying empirical beliefs, is largely a myth. For McDowell there is no such thing as "the Given" which reveals itself in experience and hence the dualism which embraces such a spurious notion is unacceptable. When I say that McDowell is gentle in his criticism what I mean is that he gives a compassionate explanation of why the empiricist tradition was so deeply impressed by the idea of "the Given" and why they had such deep faith in its justificatory properties. Davidson's coherentism was insensitive to the need that traditional empiricists felt for having a rational constraint over our empirical belief from outside, i.e. from the external world. The idea of "the Given" seemed to meet this need admirably for the traditional empiricists. McDowell is one with Davidson in acknowledging that the given can play no justificatory role as there is no such thing as "the Given." But he, unlike Davidson, and like the traditional empiricists, felt the need for having some rational constraint from the side of the world on our empirical beliefs. He thinks that in Kant we can trace the most well-rounded theory of empirical knowledge.

The critique of traditional empiricism that we find in Davidson and McDowell has a deep significance to epistemology today, especially because the debate between them has generated a dialogue between traditional and modern epistemology. What is significant is that both their criticisms are deeply rooted in the empiricist tradition which they try to criticize. Davidson did not clearly acknowledge the influences of this tradition on his thought, but McDowell acknowledges it. In their critique of the idea of the bare given they are in fact both influenced by Kant's thinking.

McDowell says that we find in Kant a picture of the generation of empirical knowledge that does not embrace the idea of "the Given." McDowell says that it is Kant who has taught us that neither "unconceptualized sensory input standing in no relation to conceptual thought ('intuitions without concepts are blind')" nor "purely intellectual thought operating independently of all rational constraint from sense experience ('thoughts without content are empty')" can help us in understanding how empirical knowledge is possible.[4] Kant tried to show that empirical knowledge is generated by intuitions involving the "faculty of receptivity" and concepts involving the "faculty of spontaneity." McDowell wants to revive this Kantian insight by showing that our conceptual capacities are operational in experiences itself and that the outer world does indeed exercise a rational check over that very experience. McDowell claims that the Kantian model of empirical knowledge does not lead us to idealism, but in fact it is the only way in which we can restore the objective world. In this discussion I would like to show in the backdrop of Davidson and McDowell's works how Kantianism with respect to empirical knowledge may be revived to exorcise the empiricist tradition of the "myth of the Given." I shall first try to state the idea of the dualism between scheme and content.

II

Empirical Knowledge — Kant and McDowell

McDowell takes Davidson's critique of the dualism as his starting point. One thing he makes clear right at the beginning. Davidson himself did not use the word "Given" while he was discussing this dualism; he rather spoke of the dualism between scheme and "content." We have to remember that it

4. M. Friedman, "Exorcising the Philosophical Tradition", in N.H. Smith (ed.), *Reading McDowell: On Mind and World*, London: Routledge, 2002.

is the concept of "Given" which is referred to as "content" when Davidson talks about scheme content dualism. "Content" is a word which is in great use in modern-day philosophy. But what we mean by the word "content" is representational content. But surely Davidson is not talking about representational content because it is conceptual and can never be dualistically opposed to what is conceptual. To understand why Davidson used the word "content" in order to mean "the Given," we have to recall Kant's remark, "thoughts without content are empty."[5] What Kant means by "content" here will be clear from the rest of the sentence quoted above, which is: "intuitions without concept are blind." John McDowell remarks, "Thought without content . . . would be a play of concept without any connection with intuition, that is, bits of experiential intake."[6] What is being suggested by Kant here is that our empirical beliefs get their content from experiential intakes.

Let us now try to understand what Kant says about the generation of empirical knowledge. Kant spoke of intuitions and concepts when he was trying to explain how we have empirical knowledge. He emphasizes on the co-operation of sensibility and understanding. Kant specifically calls this as a co-operation between "receptivity" and "spontaneity." He says:

> The receptivity of our mind, its power of receiving representation insofar as it is in any wise affected is to be entitled sensibility, then the mind's power of producing representations from itself the spontaneity of knowledge,

5. I. Kant, *Critique of Pure Reason*, tr. N.K. Smith, London: MacMillian, 1929, A51/B 75.
6. J. McDowell, *Mind and World*, Cambridge, MA: Harvard University Press, 1993, p. 4.

> should be called the understanding. Our nature is so constituted that our intuition can never be other than sensible, that is, it contains only the mode in which we are affected by objects: the faculty, on the other hand, which enables us to think of the object; of sensible intuition is the understanding. To neither of these powers may a preference be given over the other. Without sensibility no object would be given to us, without understanding no object would be thought of. Thoughts without content are empty, intuitions without concepts are blind.[7]

McDowell first of all tries to bring out why Kant uses the words spontaneity and receptivity. Kant takes the faculty of understanding as the faculty of spontaneity as he believes in a strong connection between reason and freedom. For Kant, as humans we are essentially rational. This is why when humans act from reason Kant says that they act freely or spontaneously. McDowell says that Kant's slogan would be "(T)he space of reason is the realm of freedom."[8] Sensibility on the other hand is the faculty of receptivity. But we have to remember that we are not totally free nor are we totally passive in receiving the intake from the outer world. Receptivity works together with the faculty of spontaneity in the generation of empirical knowledge.

Empiricists and realists felt that the freedom of our conceptual capacities have to be somehow tamed by the world outside. Especially if empirical thought is totally independent and free then we cannot ensure that judgements of experience are grounded in a way that would relate them to the world which is external to our thoughts. If in empirical knowledge

---

7. I. Kant, op. cit.

8. J. McDowell, op. cit., p. 5.

the world has to be presented to us then, then that very world has to exercise some kind of rational control over our conceptual capacities, otherwise we would not be able to answer the sceptics who question the very possibility of empirical knowledge. This is why philosophers felt the need of talking about the faculty of receptivity and its role in the generation of empirical knowledge. If we take these two faculties as operating separately this will give rise to "the dualism of the scheme and the given." This dualism was initially intended to ensure an external check on our freedom of employing concept in empirical knowledge. In order to find justification of our empirical thoughts we have to ground them on impingement on the conceptual realm from outside. This gives rise to the notion of "the Given."

Those who believed in the dualism did not want to underplay the role of concepts in the generation of empirical knowledge either. To start with we have to understand that when we provide empirical justification we do depend upon rational relation between concepts. The concepts which we use are empirical concepts. Observational concepts, for example, are deployed in judgements which are related to experience. It is quite obvious that the formation of such concepts would have been impossible without the intervention of "the Given." But one thing must be kept in mind, experience provides us with a "manifold of Given." It is reason which helps us to abstract the observational concept out of this manifold. So in the case of acquisition of observational concept we see that both reason and "the Given" have a part to play.

We will now try to understand why the idea of "the Given" had enamoured the empiricist tradition for so long. We have seen many foundationationalist philosophers who have established their entire theory of justification on the basis of the idea of "the Given." But the role that this "the Given" was expected to play was quite a myth.

The question is whether the "the Given" can act as a check over our rational activities. The need for suggesting "the Given" was felt in the wake of sceptical challenges brought against the possibility of empirical knowledge. It was felt by empiricists that there should be an extra-conceptual source of justification in the case of empirical knowledge. But there is a problem here: what we are getting from the non-conceptual "Given" is just the raw materials for justification. We have to somehow relate them in order to provide justification and the act of relating has to be performed by reason and by deploying conceptual abilities. As we have seen in Davidson's works, "the Given" can only justify sensations and not the beliefs that we come to have from those sensations. Philosophers found it particularly difficult to give up the idea of "the Given" and their faith in its justificatory capacity as they felt vulnerable to scepticism unless they believed in "the Given" and its justificatory role. It was felt that once we give up the notion of "the Given" the picture we will subscribe to will be "a frictionless spinning in a void."

McDowell says that we can free ourselves from an eternal see-saw between accepting and rejecting the myth of "the Given" only by adopting a view that assigns a joint role to the faculty of spontaneity and the faculty of receptivity in the generation of empirical knowledge. This can be done through a revival of Kantian insights on this matter. What we have to learn from Kant is that spontaneity and receptivity do not act separately. What we need to realize is that it is even notionally impossible for us to separate receptivity and spontaneity.

Let us now see how spontaneity and receptivity work together. For Kant intuition is nothing but experiential intake.

But what is it that we get in experience? Is it not of the form "That things are thus and so?"[9] How then can the experiential intake be non-conceptual? Even if we take our most simple experiences we will see that they are conceptualized. It is not as though our conceptual capacities are waiting for the experiential intakes over which they may be exercised to bring about empirical beliefs. Rather, our conceptual capacities are infiltrating our experiential intake. The experiential intake in that case is no longer the non-conceptual mere given. The content of experience is already conceptual. Whenever we try to justify our claims to empirical knowledge the job of justifying has to be done both by intuitions and concepts.

McDowell says that our conceptual capacities are active in experience itself. But he urges us to understand that we are not applying our concepts consciously in experience. It is just that they are already present in our experience. One might feel that this will again bring about the problem of justification. To ensure the justificatory role of "Given," it was supposed that experience should be passive. This passivity is somewhat curbed if we take "the Given" as already conceptual. One might say that since McDowell is saying that in experience concepts are simply present and not consciously applied, such a notion of concepts is strange. It is as though such concepts are not related to the faculty of spontaneity. Philosophers might think that the concepts which McDowell thinks are drawn on in experience, and are actually not concepts at all. In answer to this objection McDowell says that such an objection would have been true if it was the case that those concepts were only in play in experience. But that is not true. The very concepts which are drawn on in experience are the concepts which we use in abstract reasoning. So they are indeed genuine concepts.

---

9. J. McDowell, op. cit., p. 8.

Suppose I am experiencing a coloured computer monitor in front of me. In this experience the concepts of colour and substance are in play. These very concepts are used when we make a judgement in logic to the effect that all coloured things are extended. So these concepts are indeed concepts which result from the faculty of spontaneity.

This is how McDowell tries to demonstrate that scheme and content are not dualistically opposed as Davidson takes them to be. When Davidson is criticizing the foundationalists, he is saying that foundationalism gives experience a justificatory role by making "the Given" dualistically opposed to our conceptual capacities. He does not even consider the Kantian option where the faculty of spontaneity and receptivity are seen to be in co-operation generating empirical knowledge. This is why McDowell thinks that Davidson went wrong. Though McDowell agrees that the traditional way of seeing "the Given" will neither help us in establishing foundationalism, nor would it give us a strong theory of justification. We have to understand, according to McDowell, that the idea of "the Given" within the history of emistemology came as "a way of thinking that underlies the familiar philosophical anxiety about empirical knowledge"[10] and McDowell says that this way of thinking is also present in Davidson's own works. What led philosophers to the idea of "the Given" is the thought that our conceptual capacities in the Kantian sense belongs to the "faculty of spontaniety" and that this brings about the worry that there may be no rational check from the outside on these conceptual capacities. Consequently we are faced with the sceptical challenge that we might not have the sort of justification for our empirical knowledge.

10. J. McDowell, op. cit., p. 9.

McDowell says that Davidson belongs to that empiricist tradition where the need for an external check on our empirical beliefs is taken seriously. But what is different in Davidson is that he refuses to admit a rational constraint of our beliefs from the outside. This, McDowell says, is an "unnerving idea." McDowell feels that this would make the spontaneity of our conceptual thinking completely free and this in turn would keep the possibility of sceptical challenge open. The only way in which we can save our theory of empirical knowledge of scepticism as well as the mythical notion of "the Given" is by adhering to a Kantian view of empirical knowledge in which experiences that justify our empirical beliefs already have conceptual content.

III

McDowell on Non-conceptual Content of Perceptual Experience

Perceptual experience presents us with the most interesting case at hand. While perception is regarded as the means of knowledge that achieves the most significant bridging of the gap between the inside and the outside, it also has been the most interesting phenomenon concerning which the inner-outer distinction, its acceptance, denial, partial acceptance and partial denial have lead to the most challenging controversies in philosophy of mind and epistemology.

McDowell's particular way of handling the inner-outer distinction had a specific effect on his idea of content of perceptual experience. In the third lecture of *Mind and World* McDowell criticizes a particular idea of perceptual experience which has received great attention in recent times, viz. non-conceptual content of experience. The idea of non-conceptual content of experience was first introduced by Gareth Evans in his *Varieties of Reference*.[11] Evans does this while he discusses the role that information links play in demonstrative identification and other kind of identification. Evans in this connection speaks of the information that is resulted by the perceptual system (also in the case of somatic proprioception). He says that this information is non-conceptual and that it is initially unconscious though is made conscious when it is taken as an input to a thinking, concept-applying and reasoning system. This idea of non-conceptual content of perceptual experience was supported by all those philosophers who believed that perceptual content is "finer grained" than any of the concepts in terms of which we report our perceptual experience. "The rich detail of features presented in our experience outstrips the conceptual capacities of even the most sophisticated creatures, and this one could not possess all the concepts which are used in giving the report of the representational content of her experience."[12] There is an important debate between Christopher Peacocke and John McDowell over this issue. While Peacocke claims that our perceptual content is finer grained than any concepts that are available to us and hence such a content is bound to be non-conceptual, McDowell says that this notion of non-conceptual content of perceptual experience is unacceptable because we can never show how under such a theory of content perceptual experience can be regarded as an input that helps us to fix and justify our beliefs.

It is accepted by most people that content of propositional

11. G. Evans, *Varieties of Reference*, Oxford: Oxford University Press, 1982.

12. W. Wright, "McDowell, Demonstrative Concepts, and Nonconceptual representational Content", *Disputatio*, 14 (2003) p. 39.

attitudes are constrained by concepts. Now what kind of conceptual constraint are we talking about here? There are two ways in which this constraint is understood. Firstly, we have to remember that when we specify what the thinker thinks, what a perceiver perceives or what a speaker says by uttering a sentence on a particular occasion, we have to respect the way in which the thinker, perceiver or the speaker views or apprehends the world. Secondly, the particular way in which the speaker, perceiver or the thinker apprehends the world while speaking about the world or entertaining beliefs about the world, is actually a function of the concepts that the subject has in her repertoire. It is this second idea of constraint that follows from McDowell's brand of conceptualism.

It is this idea that has been criticized most vehemently by those who have subscribed to the idea of non-conceptual content of perceptual experience. What they say is that

> The rich detail of features presented in our experience outstrips the conceptual capacities of even the most sophisticated creatures, and thus one could not possess all the concepts which are used in giving a report of the representational content of her experience. This would seem to rule out the possibility of the representational content of experience being entirely conceptual in nature, leaving non-conceptual content, whatever it might turn out to be, as an option the representationalist must seriously consider.[13]

Let us try to understand what the non-conceptualists have in mind. Suppose a person has a perceptual experience with a particular representational content. The non-conceptualist would say that in order to have the experience the subject need not possess the concepts that are required in the specification of the representational content of the experience

13. W. Wright, op. cit., p. 39.

in question. We, while theorizing, might use certain concepts in order to state the correctness condition of the experience in question, but that does not mean that the subject herself has to possess those concepts in order that she can actually experience them. The non-conceptualists are not saying that the subject in order to have a perceptual experience with a non-conceptual representational content, say R, need not possess any of the concepts that "correspond to the features that are included in R."[14] The non-conceptualists are not even saying that if the subject were to later on come to possess those concepts then the non-conceptual content of the experience would change into conceptual content. "Non-conceptualism is a thesis about what determines the representational content of experience, and thus there is nothing barring a perceiver from possessing concepts which correspond to features included in the content of her experience. Those concepts, however, would have no role in determining the representational content of her experience." The conceptualists would deny just this. They would say that the concepts that get mentioned in the specification of the correctness condition of the experience must determine the having of the experience itself. Under the conceptualist thesis there would be significant difference in the representation content of two persons' experiences who are both seeing the same scene under the same viewing conditions if there were to be any difference in their conceptual repertoire. The non-conceptualists may differ amongst themselves. Evans for example held that the representational content of perceptual experience is completely non-conceptual. Peacocke on the other hand would say that there are both conceptual and non-conceptual properties of experience. Another view may be

14. W. Wright, op. cit., p. 40.

traced in Peacocke where the possibility of non-representational non-conceptual content of experience is conceded. We shall not go into a detailed discussion of all these varieties of non-conceptualism and what their arguments are. For our purpose we shall consider just McDowell's main argument against the non-conceptualist idea that experience is so fine-grained that our conceptual capacities cannot explain their fine-grainedness. In doing so McDowell actually questions the notion of the un-conceptualized Given and hence questions the inner-outer distinction once again. What is more interesting is that McDowell brings in perceptual demonstrative concepts in order to explain this. I find this interesting because in another place McDowell seems to be challenging the inner-outer distinction in yet another way while he speaks of perceptual demonstrative thoughts. We shall discuss this in the final section.

Let us now see what McDowell exactly says about the fine-grainedness of the content of experience. Let us look at the example that the philosophers take up to prove the fine-grainedness of experience. Any person is able to distinguish all the various shades of a colour perceptually but that does not imply we have each and every colour concept in question. We can perceptually discriminate more colours than we have colour concepts. McDowell who tries to answer this criticism against his conceptualism says that we can explain this colour experience by the help of articulable demonstrative concepts. What he would say is something like this. We manage to distinguish several shades of a colour using demonstrative concepts like that shade and this shade. If such demonstrative concepts do the job of discrimination in perceptual experiences then there would be no need for having the notion of non-conceptual content of experience. On top of that this kind of conceptualized content of experience that uses demonstrative concepts would also manage to establish how such experiences help us in fixing and justifying beliefs. Let us see how exactly demonstrative concepts help us here. McDowell says that the capacity that is associated with our demonstrative thought is such that it helps us to specify or pick out those features of our experience for which we do not have an appropriate concept. Suppose we are presented with a particular shade of red (say red_{21}) in our visual experience, as we do not have the corresponding RED_{21} concept we single out this shade as "that shade of red." This is how the fine-grained feature of my colour perception gets conceptualized and specified in language according to McDowell.

In emphasizing our capacity of demonstrative thought we should not ignore the details in which representational content of experience presents itself. McDowell is not insensitive to this. In explaining the fine-grainedness of experience McDowell says that both our capacity for demonstrative thought and the general concepts that we possess come to my assistance. To explain this McDowell says

> It is possible to acquire the concept of a shade of colour, and most of us have done so. . . . In the throes of an experience of the kind that putatively transcends one's conceptual powers — an experience that *ex hypothesi* affords a suitable sample — one can give linguistic expression to a concept that is exactly as fine-grained as the experience, by uttering a phrase like "that shade," in which the demonstrative exploits the presence of the sample.[15]

McDowell needs to show that the demonstrative abilities that he speaks of are actually playing a role in the determination of the content of experience. And McDowell says that in order

15. J. McDowell, op. cit., pp. 56-57.

that this be so we have to say that our demonstrative capacities actually persist beyond the duration of the experience and come handy in categorizing the features present in the experience in such a way that we are able to compare between past and present experiences. McDowell claims that these demonstrative capacities are recognitional. If they are not so then it would be difficult to see how the content of experience that makes use of these demonstrative capacities could be useful in the sense of being used later on in course of belief formation and belief justification.

Under this conceptualist view of content of experience the idea of the non-conceptual Given is thus challenged and the Inner-Outer Distinction brought into suspicion. What I find interesting is that McDowell while discussing the relational nature of perceptual demonstrative thought has again challenged the inner-outer distinction — but now in a different way — where the idea of the inner subjective is questioned. This anti-Cartesian approach to philosophy of mind really brings the distinction between the inner and outer and between inside and outside into question.

IV

Cartesianism, the Inner-Outer Distinction and Philosophy of Mind

One of the important contribution that McDowell makes in his *Mind and World* is developing an anti-Cartesian philosophy of mind. We shall be discussing how this anti-Cartesianism finds expression in his work on perceptual demonstrative thoughts. But before that we need to understand how the inner-outer divide that is nested in the Cartesian approach to mind and world has led to serious problems in the philosophy of mind. The inner-outer distinction was heralded in Descartes's famous *Real Distinction* between the mind and the body. This dualism of mind and body or the rest of the world makes the gulf between the mind and the world towards which it can be directed unbridgeable. McDowell's suggestion is that instead of attempting at bridging the gap between intentionality and objects we should undermine "the picture of mind that generates the Cartesian divide."[16] It is usually thought that Descartes immaterialism creates a threat, the threat of loss of the world. But such a threat is present also in materialist views. The main culprit here is neither materialism nor immaterialism but the Cartesian Real Distinction, which is nothing but the Inner-Outer distinction.[17]

I shall try to say why the inner-outer distinction that we find in form of Cartesianism in Philosophy of Mind is more fundamental than the distinction as we find it in Epistemology as creating the problem of empirical knowledge. We have to remember, as rightly pointed out by McCulloch in his article "Phenomenological Externalism,"[18] "knowledge — along with unwarranted opinion, with which it is contrasted — presupposes that we can direct our minds at, think about, the empirical world." It seems that under the Real Distinction framework of Descartes intentionality becomes impossible. McDowell writes,

> The threat which the Cartesian picture poses to our hold on the world comes out dramatically in this: that within the Cartesian picture there is a serious question about how it can be that experience, conceived from its own point of view

---

16. J. McDowell, "Singular Thought and the Extent of Inner Space", in his *Meaning, Knowledge and Reality*, Cambridge, Mass.: Harvard University Press, 1998.

17. See G. McCulloch, "Phenomenological Externalism", in N.H. Smith (ed.), *Reading McDowell*, London: Routledge, 2002., p. 123.

18. Ibid.

> is not blank or blind, but purports to be revelatory of the world we live in.[19]

On this Cartesian picture, beliefs, perceptions and other propostitional attitudes become "occupants of the same autonomous inner realm," and it is no less a problem than the problem of empirical knowledge, how occupants of such an autonomous inner realm can "have a representational bearing on the world."[20] The problem of empirical knowledge emanates from this problem of intentionality which we find within the picture of Cartesian Real Distinction. Let us see following McDowell and McCulloch why the Cartesian version of the Inner-Outer distinction, i.e. the Cartesian Real distinction cannot make room for one of the most key concepts of philosophy of mind, viz. intentionality.

McCulloch asks us to consider what he calls the *Demonic Dilemma*, which is as he put thus:

> Intentionality belongs either on the mind side of the Distinction, or the world side (it can't be on both, or the Distinction isn't Real). Then if, on the one hand, intentionality is on the mind side of the Distinction, it is all used up, signed and sealed, regardless of whether "its" world gets into the equation: hence the putative demon possibility that the "right" world never does anyway. The idea that the mind's activity (supposed thinking) is intentionally directed at something in particular is a sham: world, demon, it makes no difference to it as it is in itself, and the mind's activity simply co-exists alongside whatever is beyond in sublime indifference.

---

19. J. McDowell, "Singular Thought and the Extent of Inner Space", in his *Meaning, Knowledge and Reality*, Cambridge, Mass.: Harvard University Press, 1998, p. 243.

20. Ibid., p. 249.

> But if, on the other hand, intentionality is on the world side of the Distinction, then mind isn't as part of its nature directed at anything anyway, hence is not, in itself, directed at its world: where there is demon rather than world, there is no world-directedness either.
>
> It just doesn't help to suggest that intentionality must therefore be a relation that somehow straddles the Real Distinction. On a straightforward understanding, a relation can only obtain if the relata (mind, world) exist, so where there is demon rather than world, world-directedness does not obtain: and this is the second horn of the Dilemma.[21]

A Cartersian might suggest that the only way that we can salvage intentionality within this Real Distinction picture is by saying that the two realms (the inner and the outer) are "related by blind cause and effect." But there is a problem in thinking that cause can straddle the Distinction and yet retain it. This is so because "cause itself cannot provide the potential harmony between the mind and its world demanded by intentionality." The kind of harmony that is needed in the case of intentionality is such that it would ensure that we have thoughts about things like a tree, whenever there is a tree. And in this case the causal relation between the tree and the thought would not ensure the desired intentionality. Intentionality is richer than what is ensured by reliable causation because if it wasn't so then wherever there was causality there would be intentionality. This clearly is not the case. It is here that we must realize how important McDowell's insight is. He says that philosophy should not make the fact that our thoughts can be about the object a mystery. He says

21. G. McCulloch, op. cit., pp. 124-25.

> Although reality is independent of our thinking, it is not to be pictured as outside an outer boundary that enclosed the conceptual sphere. That things are thus and so is the conceptual content of an experience, but if the subject of the experience is not misled. That very same thing, that things are this and so, is also a perceptible fact, an aspect of the perceptible world.[22]

To see how this happens we have to look at McDowell's view of perceptual demonstrative thoughts. These thoughts, for McDowell, are such that the object towards which they are directed constitute the very identity of the thought. In McDowell's thesis about perceptual demonstrative thoughts we find an expression of the particular kind of content externalism that he espouses.

V

The Inner-Outer Distinction and McDowell on Perceptual Demonstrative Thought

Some philosophers have felt that the singular components of perceptual demonstrative thoughts have to be individuated externalistically. Suppose I wear a golden bangle and I point at it and say "this bangle is made of gold." Evans, McDowell and McCulloch would say that not only is the natural kind gold is externalistically individuated but also this bangle is. What they want us to realize is that if I were to wear a different bangle I would not be having the same demonstrative thought. And if there were no bangle I was wearing but had I hallucinated that I was indeed wearing one then I would not actually be having any demonstrative thought at all. We could extend this to the twin earth situation. Suppose both Ed and Ted say "This bangle is made of gold" and suppose further

---

22. J. McDowell, *Mind and World*, p. 26.

that Ed is in fact wearing a gold bangle whereas Ted is not wearing any bangle but is hallucinating that he is wearing an indistinguishable bangle. In this case whereas Ed is having a genuine demonstrative thought Ted is not having a genuine demonstrative thought at all. And even if Ted were not hallucinating still he would be having a different demonstrative thought altogether.

We shall now see how theories of perceptual demonstrative thoughts in the semantic tradition has influenced the views of McDowell and also philosophers like Evans and McCulloch about demonstrative thoughts. So let us say what view we might have regarding empty demonstrative expressions which we use when we take ourselves to be perceiving something. What we would be asking here is: Do Ed and Ted (in the above case) express the same proposition when they utter "This bangle is made of gold"? We will see that we get two kinds of views from Russell and Strawson.[23] Russell would say that both Ed and Ted would be expressing the same proposition though in Ed's case it would be a true proposition while in Ted's case it would be a false proposition. Strawson and Evans would disagree with Russell and say that Ed and Ted do not express the same proposition. In case of sentences involving perceptual demonstratives, they hold that the existence of the proposition that is expressed by sentences like "This bangle is made of gold" is dependent on the existence of the object referred to by the demonstrative. And they would also say that the very identity of the proposition is dependent on the identity of the object referred to by the demonstrative. This is why one could not actually be grasping the proposition unless one knew

23. We might here note that Russell and Strawson have similar differences in their opinion about sentences with empty definite descriptions.

whether there was something that the demonstrative referred and also unless one knew to what the demonstrative referred.

Let us now see what kind of arguments can be given in favour of this semantic externalism about perceptual demonstratives. We shall be concentrating on the works of Gareth Evans and John McDowell. We may just start off with two quotations from Gareth Evans and John McDowell. Evans speaks of singular thoughts as Russellian thoughts (i.e. thoughts that abide the restriction imposed by *Russell's Principle*) which he defines as "A thought is Russellian if it is such that it simply could not exist in the absence of the object or objects which it is about."[24] McDowell resonates the same idea when he says about a singular thought that it "is a thought that would not be available to be thought or expressed if the relevant object, or objects, did not exist."[25] We shall discuss the works of these philosophers on perceptual demonstrative thoughts in particular as it is in the case of such thoughts that their externalistic arguments seem to me to be most clinching and it is here that we see how the inner-outer or the inside-outside distinction or the Cartesian *Real Distinction* is undermined.

VI

Evans on Perceptual Demonstrative Thought

Let us start with Gareth Evans' view as we find it in his *Varieties of Reference.*[26] Though we shall be concentrating on Gareth Evans' *Varieties of Reference* we must mention that many of Evans' views on singular reference can be traced to his earlier

24. G. Evans, op. cit., p. 71.

25. See his "Truth-Value Gaps", reprinted in his *Meaning, Truth, and Understanding*, Cambridge, MA: Harvard University Press, 1998, p. 204.

26. G. Evans, op. cit., p. 71.

papers such as "The Causal Theories of Names." In this paper Evans distinguishes between description theory and causal theory of names, both of which receive criticism from him. He says that the reference of a name is an object which constitutes the source of information that the speaker associates with the name. He rejects the descriptivist account of object-directed thought. Evans thinks that a subject may succeed in having a thought directed to a particular object simply because she stands in a contextual relation to the object in question though she is unable to frame any description that uniquely fits the said object. Evans was not even sympathetic to causal theories of reference like that of Saul Kripke. According to Kripke, references of names are fixed in terms of causal links between the current use of the name and the event of initial baptism. Evans' critique of this view is that references of names change over time. "Madagascar" is an example of such a name which has changed its reference over time. Let us now concentrate on Evans' positive thesis. Evans tried to argue that Frege believed on what might be called the "no reference, no thought" principle. This he believed because Frege did believe that the semantic function of a name was to introduce an object. This interpretation of Frege creates a tension between itself and Frege's admission which was prompted by his famous sense-reference distinction that there could be names with sense which failed to have reference. In order to resolve this tension Evans speaks of Frege's view that empty names can have what may be called only fictional use expressing pretended senses or "mock thoughts." Evans then went on to show that especially perceptual demonstratives such as "that vase" or "that ball" confirmed to the "no reference, no thought principle." This is nothing but the version of Russell's Principle that Evans would like to adopt. And that is why this externalism is what may be called, following Jessica Brown,[27] Illusion Version of Externalism, as on this view if the desired object of the demonstrative thought is non-existent it is believed that the subject has an illusion of a thought but not a real thought.

We may now go onto the discussion of perceptual demonstrative thoughts. Evans tries to demonstrate in the sixth chapter of his *Varieties of Reference* that at least in the case of perceptual demonstrative thoughts we have to say that they are partly individuated by the objects they are about. We can take our earlier example. Suppose a subject is looking at a golden bangle b_1. She has a thought in which she thinks of the bangle in a non-descriptive mode. She thinks: this bangle is golden. Evans would say that in a counterfactual situation if the subject were to see a numerically different yet qualitatively indistinguishable bangle then she would have different thought altogether. And in the counterfactual situation where the subject is seeing no bangle at all but is having an illusion of seeing a bangle qualitatively identical with the actual one then we would have to say that the subject is having an illusion of thought. Keeping in mind Jessica Brown's classification, Evans' externalism would be at the same time the *illusion version of externalism* and *Fregean externalism*. It is Fregean because Evans claims that the thought is not only determined by the object but also by the way in which the object is presented to the subject. But this way in which the object is presented to the subject is itself also determined by the object.

Evans' particular brand of externalism can be understood if we look into his conception of the conditions that one has to satisfy in order to have a thought of an object. We have already spoken about these conditions and how they differ from

27. J. Brown, *Anti-Individualism and Knowledge*, London: MIT Press, 2004.

conditions that causal theorists have spoken of. But let us be clear about it more by bringing in Evans' own example rotating steel balls.

Suppose a person one day in the past sees two indistinguishable steel balls suspending from the same point and rotating about it. Since the subject entertains no belief about the second ball which she doesn't entertain about the first there seems to be no way in which she can distinguish one from the other. But if she were to, then she would find some distinguishing feature about them. But though this may seem intuitively plausible yet Evans points out that there may be strong intuitions against it. Let us suppose that the same subject had seen one of the balls rotating one day briefly and then on the next day she saw the other one rotating. Suppose further that the subject due to localized amnesia forgets the first episode and years later thinks about "that shiny ball" she saw many years ago. In such a situation if she was asked which ball she was thinking of she would not be able to give an answer as for her there would be one and not two balls. Would we ascribe her the thought about the second ball? If we follow the causal theory of reference and thoughts, given the amnesia of the subject we would be forced to claim that the subject's memories have causal connection with only one ball. And under the causal story we would have to say that the subject is actually having a thought about the second ball. Evans would say that even if we admit that the subject's thought is related to just one of the balls yet we do not need to admit that the subject manages to think about that ball in virtue of being causally related to it. This is so because Evans says that the subject will not be able to do anything which would demonstrate her ability to discriminate between the two balls and hence it would be wrong to ascribe to her a thought of any particular ball. One might think that Evans' desired conditions are far too difficult for any person to meet. For the time being I shall be echoing Jessica Brown's view about the conditions that Evans speaks of.[28] Brown says that had Evans said that in order to have a genuine thought about an object the subject is required to distinguish the object in question "by appearance alone. Instead, Evans allows that a subject can distinguish an object from all others by exploiting her environmental relation to it."[29] Evans says that there are three ways in which a subject might be able to distinguish an object from all other objects: by perception, by recognitional capacity and by individuating descriptions. In case of perceptual demonstrative thoughts we shall be concentrating on the distinction between perceptual demonstrative thought and recognition-based thought. Evans says that while perceiving an object a subject is able to distinguish the object from every other thing, by its specific spatial location and in terms of the type of thing it is. Perception "enables one to think about an object so long as it enables one to locate the object and discover its type, even if one has not yet done so" and this does not require the subject's being able to distinguish the object in question on the basis of appearance alone. On the other hand, Evans says that when a subject has a recognitional capacity with respect to an object she will be able to distinguish the object from all other objects by both appearance and location. In addition to this unless "a subject has some appreciation of the role of spatiotemporal considerations in the reidentification of objects" we cannot say that the subject has recognitional capacity with respect to that object. We must note that even in the case of recognition-based thoughts Evans will be talking of an externalist individuation of them.

28. G. Evans, op. cit., ch. 6; and J. Brown, op. cit., ch. 3, sec. 2-4.

29. J. Brown, op. cit., p. 79.

VII

McDowell on Perceptual Demonstrative Thought

Let us consider now John McDowell's treatment of perceptual demonstrative thoughts. At the very outset we have to keep in mind that McDowell finds the support for his variety of externalism (where he speaks of the object dependence of thoughts) from the Fregean account of thoughts. So his is a kind of Fregean Externalism, to borrow Jessica Brown's classification once more. McDowell, like Evans will emphasize on the Fregean notion of thoughts or propositions being composed of senses, which are nothing but ways of thinking about objects or modes of presentation of objects. The main support of this kind of externalism can be had from the idea that it makes no sense to talk about a way of thinking about an object where there is no object to be presented, and no sense to talk about the mode of presentation of an object where there is no object to be presented in the said mode. Consequently if a sentence, which expresses a thought or a proposition, contains a singular term which is supposed to refer to an object but does not as the object does not exist, then the thought itself would be empty just as the sense of the singular term in question would be. This however does not bring out McDowell's point about the peculiar feature that some singular terms which occur in the expression of singular thoughts which brings out the object dependence of those thoughts in a way that vindicates an externalist position. This we find in another argument. We shall now consider this argument.

Let us start off with our old example. Suppose a subject is perceiving a golden bangle in front of her and she says "This bangle is golden." On McDowell's view perceptual demonstrative thoughts such as this are ontologically dependent on the aspect of the world towards which they are directed. This ontological dependence is explained by McDowell in the following way: X will have a thought about O provided O exists and X is appropriately related to O. Here again we hear the echoes of *Russell's Principle*. McDowell explains this by saying that in order that a subject might have a genuine perceptual demonstrative thought which may be expressed as "(T)his bangle is golden," the relevant senses involved in the sentence cannot be empty. The demonstrative expression "this bangle" refers to a particular bangle that the subject sees and corresponds to a particular sense which is the way in which the bangle is presented to the subject. If there is no bangle at all then the sense in question would be empty and actually the subject would have only an illusion of thought. So here again we have an *Illusion-based Fregean Externalism*.

Let us now see what are the specific reasons for thinking that thoughts that are expressed using perceptual demonstrative terms are object-dependent. We first consider the expression of such thoughts. Following Strawson we may first of all say that the object that is referred to by a perceptual demonstrative term is essential for the determination of the semantic character of the statement of the perceptual demonstrative thought in question. The claim is to be understood in this way. In case there is no object to be referred to by the perceptual demonstrative term then there is no statement made at all and if there is a change in the object that is referred to by the demonstrative term then an altogether different statement is made. The point that McDowell would like to make is that in the absence of the object in question the statement could not be assessed as true or false and in the case where the object changes the truth conditions of the statement in question would be completely different.

Now let us extend this insight to perceptual demonstrative thoughts. The principles for perceptual demonstrative thoughts are "No object, no thought" and "Different object, different thought." Here too in the case of thoughts which have no objects corresponding to the perceptual demonstrative term embedded in the expression of that thought cannot be assessed as true or false. McDowell is unwilling to admit a third truth-value both in the case of such statements and thoughts. He prefers to say that such thoughts and statements corresponding to them would fail to be genuine thoughts and statements respectively.

Let us now try to understand what could be the most compelling reason for this bold externalism about perceptual demonstrative thoughts. Suppose I point to the golden bangle I wear and say "This bangle is made of gold." First of all let us consider what it takes for you to understand the meaning of what I say. The two concepts about which the question is asked, viz. to understand the meaning are ambiguous. In Philosophy of Language we make a distinction between the character of the sentence type and the content of the utterance made. In one sense any person who understands the English language would be able to understand the meaning of the sentence, "This bangle is made of gold." One understands the meaning of the sentence because anyone who knows English knows the rules associated with the use of such a sentence type. But there is another sense of understanding the meaning of a sentence where it is not enough for one to know just that. Unless one knows the content of the sentence one is unable to know the information conveyed by the utterance. In the case of a sentence expressing a demonstrative thought one needs to know which object is being referred to by the demonstrative in order to understand the content of the thought and the sentence that expresses it. This is so because the object is essential to the identity of the thought and also to the meaning (content) of the sentence expressing the thought. In our example, in the absence of the bangle no thought is expressed by the sentence "This bangle is made of gold" and there is no genuine thought entertained by the subject and no thought to be grasped by the hearer of the said sentence.

One may say that though the thought cannot be grasped it might nevertheless exist. Such a suggestion would be incorrect in the case of demonstrative thoughts. Let us see why this is being said. Let us take our example once again. If there is no golden bangle then there is nothing that I can make an ostension to, and there is no way in which I or anyone else can determine the truth or falsity of my statement to the effect "This bangle is made of gold." Maximillian de Gaynesford in his *John McDowell*[30] points out how one would react if I were to point at nothing and utter "This bangle is made of gold." He says that the natural response of a person on hearing such an utterance would be that of puzzlement. "But not about what I am thinking; rather about why I have expressed myself this way, how I may be deluded, and so on." One would not "suppose there is some logically incommunicable thought here" which she cannot recognize. One simply recognizes that there is no thought which can be ascribed to me and one would explain my behaviour in terms of other thoughts that she can ascribe to me. For example that I think that there is a golden bangle in front of me, but never the thought which can be expressed as "(T)his bangle is made of gold."

This treatment of perceptual demonstrative thoughts expresses the specific brand of externalism that Evans and McDowell subscribe to. Definitely this is externalism about

30. M. de Gaynesford, *John McDowell*, Cambridge: Polity Press, 2004, p. 135.

singular thoughts. The special character of their brand of externalism is that on this view the distinction between the intentional state and the object is minimized and hence the inner-outer distinction undermined.

Contributors

Sandhya Basu is former Professor of Philosophy, Rabindra Bharati University, Kolkata. She is the author of two books in epistemology, and published papers in many journals and anthologies. She has edited and co-edited a number of volumes in philosophy. She was a Fulbright Post-doctoral research fellow in the U.S. in 1982-83, and a National Fellow of the University Grants Commission in 1992-94 [*address*: Department of Philosophy, Rabindra Bharati University, Kolkata–700 050].

Kausik Bhattacharya, born on 17 November 1965, PhD., a Junior Fellow of Indian Council of Philosophical Research (1994-96), New Delhi, author of numerous articles and one book on the philosophy of Sartre, presented papers in different national and international seminars, is Reader in Philosophy in Visva-Bharati, Santiniketan, India. He is interested in metaphysical issues in philosophy of science, contemporary metaphysics, philosophy of Karl Popper, philosophy of Jean Paul Sartre and epistemology [*address*: Department of Philosophy and Religion, Visva-Bharati, Santiniketan – 731 235 (West Bengal), *e-mail: <kausic@hotmail.com> <sukau2004@yahoo.com>*].

Manjari Chakrabarty is a faculty member in the Department of Philosophy and Religion, Vishva-Bharati University, Santiniketan. She was awarded the prestigious Jawaharlal Nehru Scholarship during her doctoral work. Her research interests include Philosophy of Science (in particular, theory change and theory creation in science); Philosophy of Technology (mainly, conceptual and metaphysical issues); Phenomenology and Existentialism; Creativity Studies [*address*: Department of Philosophy and Religion, Visva-

Bharati, Santiniketan – 731 235 (West Bengal), *e-mail:* <*manjari_chakrabarty@yahoo.com*>].

Nirmalya Narayan Chakraborty teaches at the Department of Philosophy, Rabindra Bharati University, Kolkata. His publications include: *In Defense of Intrinsic Value of Nature* (2004), *Pursuit of Meaning*, (2004). He has edited *Perspectives on Radhakrishnan* (2007). He has also co-edited *Empiricism and the Two Dogmas* (2006) and *Knowledge Truth and Realism : Essays in Philosophical Analysis by Pranab Kumar Sen* (2007) [*address*: Department of Philosophy, Rabindra Bharati University, Kolkata – 700 050, *e-mail:* <*nirmalya@gmail.com*>].

Amita Chatterjee is Professor of Philosophy and Coordinator of the Centre for Cognitive Science, Jadavpur University, Kolkata. She is the author of *Understanding Vagueness* (1994), *Mental Reasoning: Experiments and Theories* jointly with Smia Sirker (2009) and editor of *Bharatiya Dharmaniti* (1998) and *Perspectives on Consciousness* (2003). She has jointly edited *Manodarshan* (2003), *Some Philosophical Issues on Indian* (2003) and two PHISPC volumes on *Philosophical Concepts Relevant to Sciences in Indian Tradition* (2006, 2008). Her areas of specializations are: Logic and Navya-Nyaya, Analytic Philosophy, Philosophy of Mind and Cognitive Science. She has published articles in national and international journals of repute and her articles have been included in important anthologies [*address*: Department of Philosophy & Centre for Cognitive Science, Jadavpur University, Kolkata – 700 032, *e-mail:* <*amita_ju@yahoo.com*>].

Milan Desai teaches at the Department of Philosophy, University of Goa. She has published papers in journals and anthologies [*address*: Department of Philosophy, Goa University, Taleigao Plateu, Goa-403 206].

Amitabha Gupta is a former Professor in the Department of Humanities and Social Sciences at IIT Bombay, Visiting Professor at University of Paris and CNRS Fellow. His areas of specialization are Mathematical Logic, Philosophy and History of Science and Cognitive Science. He has published book chapters and papers and has edited two books entitled *Proof, Computation and Agency* and

Games, Norms and Reasons published by Springer and is the Guest Editor of *JICPR's* Special Issue on *Logic and Philosophy Today* (forthcoming). He is a member Association for Symbolic Logic's Committee for Logic in East Asia; Steering Committee of the Association for Logic in India and the Council of ICPR [*address*: Indian Institute of Technology, Powai, Mumbai, *e-mail*: *<agcg503@gmail.com>*].

Chanda Gupta is a former Professor of Philosophy at Jadavpur University, Kolkata. She has been a Fullbright Scholar at Harvard University in 1985 and in 1995. She has published extensively in various journals and in anthologies. Her book *Realism vs. Realism* was first published in 1995 and subsequently republished in 2002 by Rowan and Littlefield. She has also co-edited *Essays in Social and Political Philosophy* (1989), *Philosophy of Science* (1992), *Cultural Awareness and Beyond* (1998) and *Perspectives on Education* (2009). She is presently a Senior Research Fellow of the ICPR [*address*: Department of Philosophy, Jadavpur University, Kolkata – 700 032].

Bipin Indurkhya, after being awarded a PhD spent about 12 years teaching at various universities in the US, most of which was at Boston University. Subsequently he was at Tokyo University of Agriculture and Technology, Japan for over eight years. He has been with International Institute of Information Technology, Hyderabad since 2004. His current research activities include studying and modelling creativity underlying metaphors and designing creativity-support systems, usability studies involving perceptual and cognitive aspects, and developing IT and robotics tools for assisting cognition and communication for autistic and dyslexic children. He has published several papers in journals and anthologies and has a book *Metaphor and Cognition*, to his credit [*address*: IIIT, Gachibowli, Hyderabad – 500 032, *e-mail*: *<bipin@iiit.ac.in>*].

Laxminarayan Lenka, is an Associate Professor, at North-Eastern Hill University, Shillong. He was an ICPR General Fellow at University of Hyderabad (1992-94), Lecturer in Logic and Philosophy at Jhadeswar Degree College, Orissa (1994-96), Charles Wallace India Trust Visiting Fellow at University of Liverpool (1995),

ICPR Residential Fellow (1996), SRF at IIT Madras (1998) and Lecturer at North Eastern Hill University, Shillong (1998-2006), prior to his present position. He has published a number of papers in the area of Analytic Philosophy. His area of research includes Epistemology, Philosophical Anthropology, Metapsychology, Speech Acts, Wittgenstein and Quine [*address*: Department of Philosophy, North Eastern Hill University, Shillong-793 022, *e-mail*: <*lenkanehu@yahoo.co.in*>].

Madhucchanda Sen teaches in Department of Philosophy Rabindra Bharati University, Kolkata. She has co-authored with Stan Baronett *Logic* (2009) and with Marilyn Anderson, and Pramod Nayar, *Critical Thinking, Academic Writing and Presentation Skills* (2009), and co-edited with Manidipa Sen and Nirmalya Narayan Chakraborty *Knowledge Truth and Realism : Essays in Philosophical Analysis by Pranab Kumar Sen* (2007) and with Nirmalya Narayan Chakraborty *Empiricism and the Two Dogmas* (2006). She is a trustee of Anando Trust, Kolkata [*address*: Department of Philosophy, Rabindra Bharati University, Kolkata – 700 050, *e-mail*: <*madhucchandasen@gmail.com*>].

Manidipa Sen is Associate Professor in the Centre for Philosophy, School of Social Sciences, Jawaharlal Nehru University. Her areas of interest are Philosophy of Mind and Philosophy of Language. Her publications include *Thinking about the World: An Essay in De Re Thought and the Externalist-Internalist Debate* (2008) [*address*: Centre for Philosophy, School of Social Sciences, Jawaharlal Nehru University, New Delhi – 110 067, *e-mail*: <*manidipasen@gmail.com*>].

Smita Sirker is a Lecturer of Philosophy at Jadavpur University. She is also associated as a Researcher with the Centre for Cognitive Science, Jadavpur University. Her areas of interest are moral philosophy, philosophy of mind and cognitive science. She has published articles in anthologies and journals, both national and international. She has recently co-authored a book *Mental Reasoning: Experiments and Theories* (2009) [*address*: Department of Philosophy, Jadavpur University, Kolkata – 700 032, *e-mail*: <*smita_icpr@yahoo.co.in*>].

K. Srinivas is currently Professor & Head, Department of Philosophy at Pondicherry University. His areas of interest include Analytic Philosophy, Epistemology (Indian & Western), Philosophy of Science and Modern Logic. He published a number of papers and books on the above-mentioned areas. His books include: *A.J. Ayer's Logical Positivism* (1990), *A Dictionary of Philosophy* (1993), *P.T. Raju* (2001), *A Concise Dictionary of Philosophy* (2007), and *Morality, Freedom and Social Construction* (forthcoming). Professor Srinivas has travelled widely and present papers in the International Seminars and delivered invited lectures in USA, West Indies, Germany, Austria, Thailand, and France. He was a visiting faculty at the University of Oregon (USA), Chemnitz Technological University (Germany), and Assumption University (Thailand) [*address*: Department of Philosophy, School of Humanities, Pondicherry University, Kalapet, Pondicherry – 605 014, *e-mail*: *<ragsri55@yahoo.com>*].

C.A. Tommy is an Associate Professor of Philosophy in the Department of Humanities and Social Sciences at IIT Delhi. Prior to joining IIT Delhi, he taught at IIT Kanpur, where he was an Associate Professor, and Goa University. His research and teaching interests are in the areas of Philosophy of Mind, Metaphysics, and History of Early and Late Modern Philosophy (especially the trajectories and fortunes of different forms of skepticisms) [*address*: Associate Professor of Philosophy, Department of Humanities and Social Sciences, Indian Institute of Technology, Hauz Khas, New Delhi, *e-mail*: *<catomy@gmail.com>*].

Subject Index

Author Index